12

TESI

THESES

tesi di perfezionamento in Matematica sostenuta il 10 gennaio 2008

COMMISSIONE GIUDICATRICE
Mariano Giaquinta, Presidente
Luigi Ambrosio
Camillo De Lellis
Jan Maly
Stefano Marmi
Andrea Carlo Giuseppe Mennucci

Gianluca Crippa
Dipartimento di Matematica
Università degli Studi di Parma
viale G. P. Usberti, 53/A (Campus)
43100 Parma, Italy

The Flow Associated to Weakly Differentiable Vector Fields

Gianluca Crippa

The Flow Associated to Weakly Differentiable Vector Fields

EDIZIONI
DELLA
NORMALE

ISBN: 978-88-7642-340-6

Contents

Introduction

The main topic of this thesis is the study of the *ordinary differential equation*

$$\begin{cases} \dot\gamma(t) = b(t, \gamma(t)) \\ \gamma(0) = x \end{cases} \qquad \gamma : [0, T] \to \mathbb{R}^d \qquad (1)$$

under various regularity assumptions on the *vector field*

$$b(t, x) : [0, T] \times \mathbb{R}^d \to \mathbb{R}^d \, .$$

The theory is simple and very classical in the case when b is sufficiently smooth, *i.e.* Lipschitz with respect to the spatial variable, uniformly with respect to the time variable. This is the so-called *Cauchy–Lipschitz theory*. In this case we can identify a unique *flow of the vector field b*, that is a map

$$X(t, x) : [0, T] \times \mathbb{R}^d \to \mathbb{R}^d$$

which gathers together all trajectories, in the sense that it solves

$$\begin{cases} \dfrac{\partial X}{\partial t}(t, x) = b(t, X(t, x)) \\[2mm] X(0, x) = x \, . \end{cases} \qquad (2)$$

Moreover, additional regularity of the vector field is inherited by the flow, as for instance regularity of $x \mapsto X(t, x)$.

In this classical situation there is also a strong connection with the *transport equation*, an evolutionary partial differential equation which has the form

$$\begin{cases} \partial_t u(t, x) + b(t, x) \cdot \nabla_x u(t, x) = 0 \\ u(0, x) = \bar{u}(x) \end{cases} \qquad u : [0, T] \times \mathbb{R}^d \to \mathbb{R} \, . \quad (3)$$

When $\gamma(t)$ is a solution of (1), the quantity $g(t) = u(t, \gamma(t))$ is constant with respect to time: indeed

$$\dot{g}(t) = \frac{\partial u}{\partial t}(t, \gamma(t)) + \nabla_x u(t, \gamma(t)) \cdot \dot{\gamma}(t)$$

$$= \frac{\partial u}{\partial t}(t, \gamma(t)) + b(t, \gamma(t)) \cdot \nabla_x u(t, \gamma(t)) = 0\,.$$

This means that (3) can be uniquely solved thanks to the *theory of characteristics*: the unique solution $u(t, x)$ is the transport of the initial data $\bar{u}(x)$ along solutions of (1), more precisely

$$u(t, x) = \bar{u}\big(X(t, \cdot)^{-1}(x)\big)\,.$$

Recently a big interest has grown on the extensions of this theory to situations in which the vector field b is less regular. Apart from the theoretical importance of such an extension, the main motivation comes from the study of many nonlinear partial differential equations of the mathematical physics. In various physical models of the mechanics of fluids it is essential to deal with densities or with velocity fields which are not smooth, and this corresponds to effective real world situations: for instance the turbulent behaviour of viscous fluids or the shock waves produced by a supersonic airplane.

We do not try to give here an account of the extremely wide literature on this topic, but we rather prefer to illustrate with some detail a specific case which will be presented again later on and which was a motivation for one of the main results collected in this thesis, namely Ambrosio's theorem (see Theorem 2.6.1).

The theory of *conservation laws* models situations in which the change of the amount of a physical quantity in some domain is due only to an income or an outcome of the quantity across the boundary of the domain. In the general case of several space dimensions and of a vector physical quantity these equations take the form

$$\begin{cases} \partial_t u + \mathrm{div}\,\big(F(u)\big) = 0 \\ u(0, \cdot) = \bar{u}\,, \end{cases} \tag{4}$$

where $u : [0, T] \times \mathbb{R}^d \to \mathbb{R}^k$ is the unknown and $F : \mathbb{R}^k \to M^{d \times k}$ is a given smooth map, called the flux.

The well-posedness theory for these equations is presently understood only in the scalar case $k = 1$, and this goes back to Kružkov [103], and in the one-dimensional case $d = 1$, via the Glimm scheme [97], or the front tracking method of Dafermos [74], or the more recent vanishing viscosity method of Bianchini and Bressan [34]. On the contrary, the general case $d \geq 2$ and $k \geq 2$ is presently very far from being understood.

For this reason the standard approach is to tackle some particular examples, possibly in order to get some insight on the general case. For instance, consider the case in which the nonlinearity $F(u)$ in (4) depends only on the modulus of the solution:

$$
\begin{cases}
\partial_t u + \sum_{j=1}^{d} \frac{\partial}{\partial x_j} \left(f^j(|u|)u \right) = 0 \\
u(0, \cdot) = \bar{u} .
\end{cases}
\tag{5}
$$

This is the so-called *Keyfitz and Kranzer system*, introduced in [102]. For any $j = 1, \ldots, d$ the map $f^j : \mathbb{R}^+ \to \mathbb{R}$ is assumed to be smooth. A natural heuristic strategy, implemented by Ambrosio, Bouchut and De Lellis in [15] and [11], is based on a formal decoupling of (5) in a scalar conservation law for the modulus $\rho = |u|$

$$
\partial_t \rho + \operatorname{div} \left(f(\rho)\rho \right) = 0
\tag{6}
$$

and a linear transport equation for the angular part $\theta = u/|u|$

$$
\partial_t \theta + f(\rho) \cdot \nabla \theta = 0 .
\tag{7}
$$

Therefore, it is natural to consider weak solutions u of (5) such that $\rho = |u|$ is a solution of (6) in the sense of entropies (the right notion of [103] which ensures existence, uniqueness and stability in the scalar case). These solutions of (5) are called *renormalized entropy solutions* and in [11] the authors showed the well-posedness of renormalized entropy solutions in the class of maps $u \in L^\infty([0, T] \times \mathbb{R}^d; \mathbb{R}^k)$ such that $|u| \in BV_{\text{loc}}$. In this case the Kružkov solution ρ of (6) enjoys BV_{loc} regularity, but in general it is not better, due to the creation of shocks. This means that we have to deal with the transport equation (7) in which the vector field $f(\rho)$ is BV_{loc} but not better, so that we are forced to go beyond the Cauchy–Lipschitz theory.

The theory of ordinary differential equations out of the smooth context starts with the seminal work by DiPerna and Lions [84]. We want to focus now on the three main ingredients of this approach, in order to orient the reader through the material presented in this thesis; a precise description of the content of each chapter will be made later in this Introduction.

In a quite different fashion with respect to the Cauchy-Lipschitz theory, the starting point is now the *Eulerian problem*, i.e. the well-posedness of the PDE. The first ingredient is the concept of *renormalized solution* of the transport equation. We say that a bounded distributional solution u of the transport equation

$$\partial_t u + b \cdot \nabla u = 0 \qquad (8)$$

is a renormalized solution if for every function $\beta \in C^1(\mathbb{R}; \mathbb{R})$ the identity

$$\partial_t\big(\beta(u)\big) + b \cdot \nabla\big(\beta(u)\big) = 0 \qquad (9)$$

holds in the sense of distributions. Notice that (9) holds for smooth so-lutions, by an immediate application of the chain-rule. However, when the vector field is not smooth, we cannot expect any regularity of the solutions, so that (9) is a nontrivial request when made for all bounded distributional solutions. We say that a vector field b has the *renormaliza-tion property* if all bounded distributional solutions of (8) are renormal-ized. The renormalization property asserts that nonlinear compositions of the solution are again solutions, or alternatively that the chain-rule holds in this weak context. The overall result which motivates this definition is that, if the renormalization property holds, then solutions of (3) are unique and stable.

The problem is then shifted to the validity of the renormalization prop-erty. This property is not enjoyed by all vector fields: the point is that some regularity, tipically in terms of *weak differentiability*, is needed. The second ingredient is a series of *renormalization lemmas*, which en-sure the renormalization property for vector fields satisfying various reg-ularity assumptions. The typical proof ultimately relies on the *regular-ization scheme* introduced in [84] and the regularity of the vector field comes into play when one shows that the error term in the approximation goes to zero. The two most significant results of renormalization are due to DiPerna and Lions [84] and to Ambrosio [8], who deal respectively with the Sobolev and the BV case.

The third ingredient is the possibility of transferring these results to the *Lagrangian problem*, *i.e.* to the ODE, establishing a kind of extended theory of characteristics. In this setting the right notion of solution of the ODE is that of *regular Lagrangian flow* [8], in which the "good" solution is singled out by the property that trajectories do not concentrate in small sets (this approach is slightly different from the one originally adopted in [84] to obtain results at the ODE level). The strength of this approach relies in the possibility of formulating an *abstract connection between PDE and ODE*. In this way, all the efforts made on the Eulerian side are automatically translated into well-posedness results for the regular Lagrangian flow.

We now illustrate the contribution of the author to the development of this research area. The following papers, with the exception of the first one, have been produced during the author's PhD studies in 2005–2007 at the Scuola Normale Superiore di Pisa, with a co-supervision at the Universität Zürich.

In a joint work with Luigi Ambrosio and Stefania Maniglia [14] we study some fine properties of normal traces of vector fields with measure divergence and we apply them to show the renormalization property for special vector fields with bounded deformation. Moreover, we prove approximate continuity properties of solutions of transport equations. This work is not presented in detail in this thesis, since it has already been presented in the Tesi di Laurea [64]. We mention it in Section 2.7 for the results relative to the renormalization property and in Section 6.7 for some more recent progresses based on the approximate continuity results.

The paper with Camillo De Lellis [65] settles, in the negative, a conjecture made by Bressan about the possibility of applying transport equation techniques to the study of multi-dimensional systems of conservation laws, in the same spirit of the derivation mentioned for the Keyfitz and Kranzer system. This is briefly presented in Section 5.2.

In the two-dimensional divergence-free case the presence of a Hamiltonian function which is (at least formally) conserved along the flow gives an additional structure to the problem and this allows a dramatic reduction of the regularity needed for the uniqueness. In a first paper with Ferruccio Colombini and Jeffrey Rauch [57] we show that all the results contained in the literature on this problem hold also in the case of bounded, and not necessarily zero, divergence. However, a technical assumption of unclear meaning is present in all these results. A further investigation, in a work in progress in collaboration with Giovanni Alberti and Stefano Bianchini [5], identifies the sharpest hypothesis needed for the uniqueness and clarifies its meaning. A new technique of dimensional reduction via a splitting on the level sets of the Hamiltonian is introduced. The positive result, which is presented in a basic case in Chapter 4, is complemented by two counterexamples which show the necessity of the "extra assumption".

In a joint work with François Bouchut [37] the meaning of the renormalization property is discussed: its connections with forward and backward uniqueness, strong continuity and smooth approximations of the solution are analysed. Chapter 3 is completely devoted to the presentation of these results.

A second paper in collaboration with Camillo De Lellis [66] (see also the conference proceedings [67]) introduces a new approach to the theory of regular Lagrangian flows, in the $W^{1,p}$ case with $p > 1$. This approach is based on some a priori quantitative estimates, which can be derived directly at the ODE level, with no mention to the PDE theory. These estimates allow to recover in a direct way various known well-posedness results, but also have some new interesting corollaries, in terms of regularity and compactness of the regular Lagrangian flow. This work is presented in Chapter 7.

Eventually we mention the lecture notes [13] in collaboration with Luigi Ambrosio: they present a self-contained survey of the theory of regular Lagrangian flows, with a particular interest to the recent developments.

We finally illustrate the content of the various chapters of this thesis. For more details the reader is referred to the introduction of each chapter.

In Chapter 1 we present the Cauchy–Lipschitz theory: we start with the classical Picard–Lindelöf theorem regarding existence and uniqueness for ODEs with Lipschitz vector field and we present some variants and extensions. Then we prove existence, uniqueness and regularity of the classical flow of a vector field. Finally we present some ideas of the theory of characteristics and the connection between the ODE and the transport and the continuity equations in the smooth case.

Starting from Chapter 2 we begin to consider the non-smooth case from the Eulerian viewpoint. We first introduce heuristically the motivation for the concept of renormalized solution. Then we define rigorously the notion of weak (distributional) solution and we see that existence holds under very general hypotheses. The renormalization property is then introduced, together with its consequences for the well-posedness of the PDE. We describe the DiPerna–Lions regularization scheme and the proof of the renormalization property by DiPerna and Lions in the Sobolev case and by Ambrosio in the BV case. Finally we comment on other various renormalization results and on the case of nearly incompressible vector fields.

Chapter 3 is devoted to [37]: we show, first in the divergence-free framework and then in more general cases, the relations between the notions of renormalization, strong continuity of the solution with respect to the time, forward and backward uniqueness in the Cauchy problem and approximation (in the sense of the norm of the graph of the transport operator) of the solution with smooth maps.

The two-dimensional case is analysed in detail in Chapter 4, in which we mainly follow [5]. In the basic case of an autonomous bounded divergence-free vector field we show that, assuming a weak Sard property on the Hamiltonian, the two-dimensional PDE can be split on the level sets of the Hamiltonian. Then we show uniqueness "line by line", using explicit arguments and a careful study of the structure of the level sets, and this eventually implies uniqueness for the full two-dimensional equation.

Chapter 5 collects many counterexamples to the uniqueness, in particular the examples of [82] and [84] which shed light on the necessity of the assumptions made. Moreover, some related constructions are presented, in particular the counterexample of [65], which shows the impossibility of building a theory for multi-dimensional systems of conservation laws based on transport equations, and the examples of [62], relative to the lack of propagation of regularity in the transport equation.

With Chapter 6 we start to present the Lagrangian viewpoint. First, the equivalence between pointwise uniqueness for the ODE and uniqueness of positive measure-valued solutions of the continuity equation is shown, exploiting the superposition principle for solutions of the PDE. Then, we introduce the concept of regular Lagrangian flow and we show how the uniqueness of bounded solutions of the continuity equation implies existence and uniqueness of the regular Lagrangian flow. We finally comment on the DiPerna–Lions' notion of flow and on the case of nearly incompressible vector fields, with particular attention to a compactness conjecture made by Bressan.

Chapter 7 is devoted to the a priori quantitative estimates for Sobolev vector fields [66]. The estimates are derived first in the bounded case, and then assuming more general growth conditions. The consequences regarding existence, uniqueness, stability (in a quantitative form), compactness and regularity of the regular Lagrangian flow are presented. Some applications to the propagation of a mild regularity in the transport equation and to a mixing conjecture made by Bressan are also addressed.

Finally, in the Appendix we recall some basic tools and results which are widely used throughout the whole thesis.

As a general warning to the reader we observe that in many cases we do not state our results under the sharpest possible assumptions: this is done in order to simplify the exposition and to avoid annoying technicalities. For instance, in almost all the thesis we deal only with globally bounded vector fields, but more general growth conditions would be sufficient, compare Section 7.3. All the statements are typically completely rigorous, while concerning the proofs we have sometimes chosen to skip some details, to merely specify the main steps or to enlighten some important or delicate points, at the expense of a more complete proof. This is done according to the author's taste and not to the absolute importance of each result. For the main notation used we refer the reader to the Appendix or to the first place where each symbol or locution appears.

Chapter 1
The theory in the smooth framework

In this chapter we illustrate the main results of the theory of ordinary differential equations in the smooth framework. Most of these results are very classical, hence we give only a sketch of the proofs. In Sections 1.1 and 1.2 we study existence, uniqueness and stability properties of the solutions. In Section 1.3 we introduce the concept of *flow* of a vector field. In Section 1.4 we illustrate the link between the ordinary differential equation and two partial differential equations, namely the *transport equation* and the *continuity equation*. Basic references for these topics are for instance [18, 39, 94, 98, 121].

1.1. The ordinary differential equation

Let $b : D \subset \mathbb{R}_t \times \mathbb{R}_x^d \to \mathbb{R}^d$ be a continuous time-dependent vector field, where D is an open set. A (classical) solution of the *ordinary differential equation*

$$\dot{\gamma}(t) = b(t, \gamma(t)) \tag{1.1}$$

consists of an interval $[t_1, t_2] \subset \mathbb{R}$ and a function $\gamma \in C^1([t_1, t_2]; \mathbb{R}^d)$ which satisfies (1.1) for every $t \in [t_1, t_2]$. In particular this implies that $(t, \gamma(t)) \in D$ for every $t \in [t_1, t_2]$. We also say that γ is an *integral curve* or a *characteristic curve* of the vector field b. Notice that, if $b \in C^\infty(D; \mathbb{R}^d)$, then every characteristic curve of b is in $C^\infty([t_1, t_2]; \mathbb{R}^d)$.

Now we fix $(t_0, x_0) \in D$ and we consider the *Cauchy problem* for the ordinary differential equation:

$$\begin{cases} \dot{\gamma}(t) = b(t, \gamma(t)) \\ \gamma(t_0) = x_0 \, . \end{cases} \tag{1.2}$$

A solution to this problem is a function $\gamma \in C^1([t_1, t_2]; \mathbb{R}^d)$ which is a characteristic curve of b and satisfies the condition $\gamma(t_0) = x_0$. In particular we must require $t_0 \in [t_1, t_2]$.

It is immediate to check that $\gamma \in C^1([t_0 - r, t_0 + r]; \mathbb{R}^d)$ is a solution of (1.2) if and only if $\gamma \in C^0([t_0 - r, t_0 + r]; \mathbb{R}^d)$ and satisfies the identity

$$\gamma(t) = x_0 + \int_{t_0}^{t} b(s, \gamma(s)) \, ds \qquad \text{for every } t \in [t_0 - r, t_0 + r].$$

1.2. Existence and uniqueness in the classical setting

We want to discuss the well-posedness of (1.2) under the assumption that the vector field b is sufficiently smooth. The following simple example shows that the situation could be complicated even in the case of quite simple one-dimensional vector fields.

Example 1.2.1 (The square root example). We consider on $\mathbb{R}$ the (continuous but not Lipschitz) vector field $b(t, x) = \sqrt{|x|}$. The ordinary differential equation

$$\begin{cases} \dot{\gamma}(t) = \sqrt{|\gamma(t)|} \\ \gamma(0) = 0 \end{cases}$$

has the solution

$$\gamma^c(t) = \begin{cases} 0 & \text{if } t \leq c \\ \frac{1}{4}(t - c)^2 & \text{if } t \geq c \end{cases}$$

for every value of the parameter $c \in [0, +\infty]$. The solution can "stay at rest" in the origin for an arbitrary time. This example will be considered again later on: see Remark 6.3.3.

This example stresses that some smoothness assumptions are necessary to obtain uniqueness. The following classical result by Picard–Lindelöf requires Lipschitz continuity of b with respect to the spatial variable, with some uniformity with respect to the time.

Theorem 1.2.2 (Picard–Lindelöf). *Let b be a continuous vector field defined on an open set containing the rectangle*

$$D = \left\{ (t, x) \in \mathbb{R} \times \mathbb{R}^d \ : \ |t - t_0| \leq \alpha, \ |x - x_0| \leq \beta \right\}.$$

Assume that b is Lipschitz continuous with respect to the spatial variable, uniformly with respect to the time, in D, i.e.

$$|b(t, x) - b(t, y)| \leq K|x - y| \qquad \text{for every } (t, x), (t, y) \in D, \quad (1.3)$$

and let M be such that $|b(t, x)| \leq M$ in D. Then there exists a unique solution $\gamma \in C^1([t_0 - r, t_0 + r]; \mathbb{R}^d)$ to (1.2), where

$$r < \min \left\{ \alpha, \frac{\beta}{M}, \frac{1}{K} \right\}.$$

Proof. Choosing r as in the statement we define $I_r = [t_0 - r, t_0 + r]$. We will show that there exists a unique $\gamma \in C^0(I_r; \mathbb{R}^d)$ such that

$$\gamma(t) = x_0 + \int_{t_0}^t b(s, \gamma(s)) \, ds \qquad \text{for every } t \in I_r.$$

We want to identify a complete metric space X on which the operator

$$T[\gamma](t) = x_0 + \int_{t_0}^t b(s, \gamma(s)) \, ds$$

is a contraction. We set

$$X = \left\{ \gamma \in C^0(I_r; \mathbb{R}^d) : \gamma(t_0) = x_0 \text{ and } |\gamma(t) - x_0| \leq \beta \text{ for every } t \in I_r \right\}.$$

X is a closed subset of $C^0(I_r; \mathbb{R}^d)$ equipped with the norm of the uniform convergence, hence it is a complete metric space. If $\gamma \in X$ then clearly $T[\gamma]$ is a continuous map which satisfies $T[\gamma](t_0) = x_0$. Moreover

$$|T[\gamma](t) - x_0| \leq \left| \int_{t_0}^t |b(s, \gamma(s))| \, ds \right| \leq M |t - t_0| \leq M r < \beta.$$

This implies that T takes value in X.

We now check that T is a contraction. Take γ_1 and γ_2 in X. Then

$$|T[\gamma_1](t) - T[\gamma_2](t)| \leq \left| \int_{t_0}^t |b(s, \gamma_1(s)) - b(s, \gamma_2(s))| \, ds \right|$$

$$\leq K \left| \int_{t_0}^t |\gamma_1(s) - \gamma_2(s)| \, ds \right|$$

$$\leq K |t - t_0| \|\gamma_1 - \gamma_2\|_{L^\infty(I_r)}$$

$$\leq K r \|\gamma_1 - \gamma_2\|_{L^\infty(I_r)}.$$

This implies that

$$\|T[\gamma_1] - T[\gamma_2]\|_{L^\infty(I_r)} \leq K r \|\gamma_1 - \gamma_2\|_{L^\infty(I_r)}$$

and since $Kr < 1$ we can apply the Banach fixed point theorem to get the existence of a unique fixed point for T. This is precisely the unique solution to (1.2). $\qquad\square$

Remark 1.2.3. This result is also valid if we assume that b is only summable with respect to the time variable and Lipschitz continuous in x uniformly in t. For a proof see for instance [22, Theorem 4.0.7].

If we drop the Lipschitz continuity assumption, then we lose uniqueness of the solution: compare with Example 1.2.1. However, we still have existence, thanks to the following result.

Theorem 1.2.4 (Peano). *Let b be a continuous and bounded vector field defined on an open set containing the rectangle*

$$D = \left\{ (t, x) \in \mathbb{R} \times \mathbb{R}^d \ : \ |t - t_0| \leq \alpha, \ |x - x_0| \leq \beta \right\}.$$

Then there exists a local solution to (1.2).

Proof. We take $r < \min\{\alpha, \beta/M\}$ and consider again the operator T and the space X defined in the proof of Theorem 1.2.2. X is a non-empty bounded convex closed subset of $C^0(I_r; \mathbb{R}^d)$ equipped with the norm of the uniform convergence and $T : X \to X$ is a continuous operator (because of the uniform continuity of b). Moreover for every $\gamma \in X$ we can estimate

$$|T[\gamma](t) - T[\gamma](t')| = \left| \int_{t'}^{t} |b(s, \gamma(s))| \, ds \right| \leq M|t - t'|.$$

Hence by the Ascoli–Arzelà theorem T is a compact operator. Then the existence of a fixed point in X is ensured by the Caccioppoli–Schauder fixed point theorem (see [95, Corollary 11.2]). □

Remark 1.2.5. In the one-dimensional autonomous case we have (local) uniqueness provided b is continuous and $b(x_0) \neq 0$. Indeed, set $G(x) = \int_{x_0}^{x} \frac{dy}{b(y)}$. Since $b(x_0) \neq 0$ we can find a neighbourhood of x_0 in which G is of class C^1 with $G' \neq 0$; hence G has a C^1 inverse. If γ is a solution of (1.2) defined in a neighbourhood of 0 we can compute

$$\frac{d}{dt} G(\gamma(t)) = \frac{\dot{\gamma}(t)}{b(\gamma(t))} = 1,$$

which eventually implies $G(\gamma(t)) = t$ and thus $\gamma(t) = G^{-1}(t)$. Since G only depends on b, this means that the solution is unique. This is only true in a neighbourhood of $t = t_0$: similarly to the case described in Example 1.2.1, the solution of the problem

$$\begin{cases} \dot{\gamma}(t) = \sqrt{|\gamma(t)|} \\ \gamma(0) = x_0 < 0 \end{cases}$$

reaches the origin (and up to that time it is unique), but it can stay at rest there for an arbitrary amount of time before leaving it.

We now present two other conditions which are a bit more general than the Lipschitz condition required in Theorem 1.2.2, but nevertheless they are sufficient to get uniqueness.

Proposition 1.2.6 (One-sided Lipschitz condition).   *Uniqueness forward in time for* (1.2) *holds if the Lipschitz continuity condition* (1.3) *in Theorem* 1.2.2 *is replaced by the following* one-sided Lipschitz condition:

$$\big(b(t, x) - b(t, y)\big) \cdot (x - y) \le K|x - y|^2 \qquad \textit{for every } (t, x), (t, y) \in D.$$

By "forward in time" we mean the following: two solutions γ_1 and γ_2 to (1.2), which by definition satisfy $\gamma_1(t_0) = \gamma_2(t_0) = x_0$, coincide for $t > t_0$. Observe that the one-sided Lipschitz condition cannot guarantee uniqueness backward in time, since it is not invariant under an inversion of the sign of t in the equation, which amounts to a change of sign of b.

Proof of Proposition 1.2.6.   Consider $r(t) = |\gamma_1(t) - \gamma_2(t)|^2$ and notice that $r(t_0) = 0$. Using the one-sided Lipschitz condition we can estimate

$$\dot{r}(t) = 2\big(\gamma_1(t) - \gamma_2(t)\big) \cdot \big(b(t, \gamma_1(t)) - b(t, \gamma_2(t))\big)$$
$$\le K|\gamma_1(t) - \gamma_2(t)|^2 = Kr(t).$$

This implies that

$$\frac{d}{dt}\Big[r(t)e^{-Kt}\Big] \le 0,$$

hence

$$r(t)e^{-Kt} \le r(t_0)e^{-Kt_0} = 0 \qquad \text{for every } t \ge t_0,$$

that is the desired thesis.   □

Proposition 1.2.7 (Osgood condition).   *Uniqueness for* (1.2) *holds if the Lipschitz continuity condition* (1.3) *in Theorem* 1.2.2 *is replaced by the following* Osgood *condition:*

$$|b(t, x) - b(t, y)| \le \omega(|x - y|) \qquad \textit{for every } (t, x), (t, y) \in D,$$

where $\omega : \mathbb{R}^+ \to \mathbb{R}^+$ *is an increasing function satisfying* $\omega(0) = 0$, $\omega(z) > 0$ *for every* $z > 0$ *and*

$$\int_0^1 \frac{1}{\omega(z)}\, dz = +\infty.$$

Proof. As in the proof on the previous proposition we define $r(t) = |\gamma_1(t) - \gamma_2(t)|^2$. We have $r(t_0) = 0$ and we assume by contradiction that $r(t) > 0$ for $t \in]t_0, t_0 + r[$ for some $r > 0$. This local forward uniqueness clearly implies global forward uniqueness, and backward global uniqueness is proved in the same way. We compute

$$\dot{r}(t) = 2\big(\gamma_1(t) - \gamma_2(t)\big) \cdot \big(\dot{\gamma}_1(t) - \dot{\gamma}_2(t)\big)$$
$$= 2\big(\gamma_1(t) - \gamma_2(t)\big) \cdot \big(b(t, \gamma_1(t)) - b(t, \gamma_2(t))\big).$$

Using the Osgood condition and integrating with respect to the time we get

$$r(t) \leq 2 \int_{t_0}^t \sqrt{r(s)}\, \omega(r(s))\, ds. \tag{1.4}$$

We notice that $\sqrt{r(s)} = |\gamma_1(t) - \gamma_2(t)| \in L^1([t_0, t_0 + r])$ and we apply an extension of the Gronwall lemma, which goes as follows. Define

$$R(t) = 2 \int_{t_0}^t \sqrt{r(s)}\, \omega(r(s))\, ds.$$

Since we are assuming that $r(t) > 0$ for $t \in]t_0, t_0 + r[$ we deduce that $R(t) > 0$ for $t \in]t_0, t_0 + r[$. Moreover applying (1.4) we have

$$\dot{R}(t) = \sqrt{r(t)}\, \omega(r(t)) \leq \sqrt{r(t)}\, \omega(R(t)),$$

from which we get

$$\frac{\dot{R}(t)}{\omega(R(t))} \leq \sqrt{r(t)} \qquad t > t_0. \tag{1.5}$$

We set

$$\Omega(z) = \int_z^1 \frac{1}{\omega(r)}\, dr$$

and we observe that $\Omega(z)$ is well-defined for every $z > 0$, satisfies $\Omega'(z) = -1/\omega(z)$ and $\Omega(z) \uparrow +\infty$ as $z \downarrow 0$, because of the Osgood condition. Integrating (1.5) we deduce that for every $t_0 < s < t$ we have

$$-\Omega(R(t)) + \Omega(R(s)) \leq \int_s^t \sqrt{r(\tau)}\, d\tau,$$

which however cannot hold for s very close to t_0. Indeed, since $R(s) \downarrow 0$ as $s \downarrow t_0$, in this case we would have $\Omega(R(s)) \uparrow +\infty$ as $s \downarrow t_0$, while the integral on the right hand side stays finite. $\square$

We conclude this section with a discussion about the maximal interval of existence of the solution to (1.2). The solution that we constructed in the previous theorems is in fact local in time. However, if $b : D \subset \mathbb{R} \times \mathbb{R}^d \to \mathbb{R}^d$ is continuous and bounded, then every solution $\gamma :]t_1, t_2[\to \mathbb{R}^d$ of (1.2) can be extended to the closed interval $[t_1, t_2]$. Indeed, for every $t_1 < t < t' < t_2$ we have

$$|\gamma(t') - \gamma(t)| \leq \int_t^{t'} |b(s, \gamma(s))|\, ds \leq M|t' - t|,$$

where M is an upper bound for $|b|$ on D. Hence γ is Lipschitz and admits a unique extension to the closure of its domain of definition.

Now notice that, if $(t_1, \gamma(t_1))$ or $(t_2, \gamma(t_2))$ is not on the boundary of D, we can apply again the local existence result proved in Theorem 1.2.2 (or in Theorem 1.2.4). We conclude the following result:

Theorem 1.2.8. *Let* $b : D \subset \mathbb{R} \times \mathbb{R}^d \to \mathbb{R}^d$ *be a continuous and bounded vector field. Assume that b is Lipschitz continuous with respect to the spatial variable, uniformly with respect to the time, on every bounded rectangle contained in D. Fix $(t_0, x_0) \in D$. Then there exists a unique solution to (1.2) and it can be extended until its graph touches the boundary of D. This identifies a* maximal interval of existence *for the solution to (1.2).*

If b is globally defined and bounded we deduce

Corollary 1.2.9. *Let* $b : I \times \mathbb{R}^d \to \mathbb{R}^d$ *be a continuous and bounded vector field, where $I \subset \mathbb{R}$ is an interval. Assume that b is locally Lipschitz continuous with respect to the spatial variable, uniformly with respect to the time. Then, for every $(t_0, x_0) \in I \times \mathbb{R}^d$ there exists a unique solution to (1.2) which is defined for $t \in I$.*

1.3. The classical flow of a vector field

We want to compare two solutions γ and $\widehat{\gamma}$ to the ordinary differential equation, with initial data at time t_0 respectively equal to x_0 and $\widehat{x_0}$. We can compute

$$\gamma(t) - \widehat{\gamma}(t) = x_0 - \widehat{x_0} + \int_{t_0}^t b(s, \gamma(s))\, ds - \int_{t_0}^t b(s, \widehat{\gamma}(s))\, ds$$

$$= x_0 - \widehat{x_0} + \int_{t_0}^t \left[b(s, \gamma(s)) - b(s, \widehat{\gamma}(s)) \right] ds \,.$$

Hence we deduce

$$|\gamma(t) - \widehat{\gamma}(t)| \le |x_0 - \widehat{x_0}| + \mathrm{Lip}(b) \int_{t_0}^{t} |\gamma(s) - \widehat{\gamma}(s)| \, ds \,,$$

where $\mathrm{Lip}(b)$ denotes, as usual, the Lipschitz constant of b. A simple Gronwall argument then gives

$$|\gamma(t) - \widehat{\gamma}(t)| \le |x_0 - \widehat{x_0}| \exp\left(K|t - t_0|\right). \tag{1.6}$$

This means that the solution depends Lipschitz continuously on the initial data.

A very similar argument shows that, if we consider two solutions γ and $\widehat{\gamma}$ with the same initial data x_0 but relative to different vector fields b and $\widehat{b}$, we have the continuity estimate

$$|\gamma(t) - \widehat{\gamma}(t)| \le |t - t_0| \|b - \widehat{b}\|_\infty \exp\left(K|t - t_0|\right). \tag{1.7}$$

Looking at the solution of (1.2) as a function not only of the time but also of the initial point x_0 we are led to the following definition:

Definition 1.3.1 (Classical flow of a vector field). Consider a continuous and bounded vector field $b : I \times \mathbb{R}^d \to \mathbb{R}^d$, where $I \subset \mathbb{R}$ is an interval, and let $t_0 \in I$. The *(classical) flow of the vector field* b starting at time t_0 is a map

$$X(t, x) : I \times \mathbb{R}^d \to \mathbb{R}^d$$

which satisfies

$$\begin{cases} \dfrac{\partial X}{\partial t}(t, x) = b(t, X(t, x)) \\[2mm] X(t_0, x) = x \,. \end{cases} \tag{1.8}$$

If the vector field b is bounded and Lipschitz with respect to the space variable we immediately deduce existence and uniqueness of the flow from Corollary 1.2.9. Moreover (1.6) also gives Lipschitz regularity of the flow.

Corollary 1.3.2. *Let $b : I \times \mathbb{R}^d \to \mathbb{R}^d$ be a continuous and bounded vector field, where $I \subset \mathbb{R}$ is an interval. Assume that b is locally Lipschitz continuous with respect to the spatial variable, uniformly with respect to the time. Then, for every $t_0 \in I$ there exists a unique classical flow of b starting at time t_0. Moreover, the flow is Lipschitz continuous with respect to t and x.*

Now we want to show that additional smoothness of b implies that the flow is also smooth with respect to the spatial variable. Assume that b is C^1 with respect to the spatial variable, uniformly with respect to the time. We first discuss the differentiability in the direction given by a unit vector $e \in \mathbb{S}^{d-1}$. For every small $h \in \mathbb{R}$ we need to compare $X(t, x)$ and $X(t, x+he)$. We first observe that differentiating formally (1.8) with respect to x in the direction e we obtain the following ordinary differential equation for $D_x X(t, x)e$:

$$\frac{\partial}{\partial t} D_x X(t, x)e = (D_x b)(t, X(t, x)) D_x X(t, x)e .$$

Motivated by this, we define $w_e(t, x)$ to be the solution of

$$\begin{cases} \dfrac{\partial w_e}{\partial t}(t, x) = (D_x b)(t, X(t, x)) w_e(t, x) \\ w_e(t_0, x) = e . \end{cases} \tag{1.9}$$

This is a linear ordinary differential equation which depends on the parameter $x \in \mathbb{R}^d$. It is readily checked that for every $x \in \mathbb{R}^d$ there exists a unique solution $w_e(t, x)$ defined for $t \in I$. Moreover it is simple to prove that $w_e(t, x)$ depends continuously on the parameter $x \in \mathbb{R}^d$.

We claim that

$$\frac{X(t, x + he) - X(t, x)}{h} \to w_e(t, x) \qquad \text{as } h \to 0. \tag{1.10}$$

This will give $D_x X(t, x)e = w_e(t, x)$ and, since $w_e(t, x)$ is continuous in x, we will get that the flow $X(t, x)$ is differentiable with respect to x with continuous differential. Going back to (1.9) we also deduce that, if b is C^k with respect to the spatial variable, then the flow X is C^k with respect to x.

We now prove (1.10). We define the map

$$z_{e,h}(t, x) = \frac{X(t, x + he) - X(t, x)}{h} .$$

We notice that $z_{e,h}(0, x) = e$ and we compute

$$\begin{aligned} \frac{\partial z_{e,h}}{\partial t}(t, x) &= \frac{1}{h} \left[\frac{\partial X}{\partial t}(t, x + he) - \frac{\partial X}{\partial t}(t, x) \right] \\ &= \frac{1}{h} \left[b(t, X(t, x+he)) - b(t, X(t, x)) \right] \\ &= \left(\int_0^1 (D_x b)\Big(t, \tau X(t, x+he) + (1-\tau) X(t, x)\Big) d\tau \right) z_{e,h}(t, x) \\ &= (D_x b)(t, X(t, x)) z_{e,h}(t, x) + \Psi_{e,h}(t, x) z_{e,h}(t, x), \end{aligned}$$

where we have set

$$\Psi_{e,h}(t, x) = \int_0^1 \Big[(D_x b)(t, \tau X(t, x + he)$$

$$+ (1 - \tau)X(t, x)) - (D_x b)(t, X(t, x)) \Big] d\tau \,.$$

Since we are assuming that the vector field b is C^1 with respect to the spatial variable we deduce that

$$\Psi_{e,h}(t, x) \to 0 \qquad \text{as } h \to 0, \text{ uniformly in } t \text{ and } x.$$

We consider $\varphi_{e,h}(t, x) = z_{e,h}(t, x) - w_e(t, x)$ and we notice that $\varphi_{e,h}$ satisfies

$$\begin{cases} \dfrac{\partial \varphi_{e,h}}{\partial t}(t, x) = (D_x b)(t, X(t, x))\varphi_{e,h}(t, x) + \Psi_{e,h}(t, x)z_{e,h}(t, x) \\ \varphi_{e,h}(0, x) = 0 \,. \end{cases}$$

Recalling (1.7) and the fact that $|\Psi_{e,h}(t, x)| = o(1)$ and $|z_{e,h}(t, x)| = O(1)$ we deduce that $|\varphi_{e,h}(t, x)| = o(1)$. Going back to the definitions of $\varphi_{e,h}$ and of $z_{e,h}$ we see that this is precisely (1.10). We conclude the following theorem, improving the result of Corollary 1.3.2.

Theorem 1.3.3. *Let $b : I \times \mathbb{R}^d \to \mathbb{R}^d$ be a smooth and bounded vector field, where $I \subset \mathbb{R}$ is an interval. Then for every $t_0 \in I$ there exists a unique classical flow of b starting at time t_0, which is smooth with respect to t and x.*

Notice also that, for every $t \in I$, the map

$$X(t, \cdot) : \mathbb{R}^d \to \mathbb{R}^d$$

is a smooth diffeomorphism. The Jacobian $J(t, x) = \det(\nabla_x X(t, x))$ satisfies the equation

$$\frac{\partial J}{\partial t}(t, x) = (\operatorname{div} b)(t, X(t, x))J(t, x) \,, \tag{1.11}$$

which also implies that $J(t, x) > 0$ for every $t \in I$. Moreover, using the notation $X(t, s, x)$ for the flow of b starting at time $s \in I$, the following *semi-group property* holds, as a consequence of the uniqueness of the flow:

$$X(t_2, t_0, x) = X(t_2, t_1, X(t_1, t_0, x)) \qquad \text{for every } t_0, t_1, t_2 \in I. \tag{1.12}$$

1.4. The transport equation and the continuity equation

In this section we show the relation between the ordinary differential equation and the *transport equation*

$$\partial_t u(t, x) + b(t, x) \cdot \nabla u(t, x) = 0, \quad u(t, x) : [0, T] \times \mathbb{R}^d \to \mathbb{R}, \quad (1.13)$$

and the *continuity equation*

$$\partial_t \mu + \operatorname{div}(b\mu) = 0, \qquad (1.14)$$

where $\mu = \{\mu_t\}_{t \in [0,T]}$ is a family of locally finite signed measures on $\mathbb{R}^d$, which depends on the time parameter $t \in [0, T]$. The continuity equation is intended in distributional sense, according to the following definition.

Definition 1.4.1. A family $\mu = \{\mu_t\}_{t \in [0,T]}$ of locally finite signed measures on $\mathbb{R}^d$ is a solution of the continuity equation (1.14) if

$$\int_0^T \int_{\mathbb{R}^d} \left[\partial_t \varphi(t, x) + b(t, x) \cdot \nabla \varphi(t, x) \right] d\mu_t(x) \, dt = 0$$

for every test function $\varphi \in C_c^\infty(]0, T[\times \mathbb{R}^d)$.

We recall here only the main ideas of the *theory of characteristics* for partial differential equations and we derive some explicit formulas for the solution in the smooth case. The connection between the *Lagrangian problem* (ODE) and the *Eulerian problem* (PDE) will be investigated in detail in Chapter 6, in which a kind of extension of the theory of characteristics to the non-smooth case will be described.

If we consider a characteristic curve $\gamma(t)$ of the vector field $b(t, x)$ and a smooth solution $u(t, x)$ of (1.13) we can easily check that the quantity $g(t) = u(t, \gamma(t))$ is constant with respect to time. Indeed

$$\dot{g}(t) = \frac{\partial u}{\partial t}(t, \gamma(t)) + \nabla_x u(t, \gamma(t)) \cdot \dot{\gamma}(t)$$

$$= \frac{\partial u}{\partial t}(t, \gamma(t)) + b(t, \gamma(t)) \cdot \nabla_x u(t, \gamma(t)) = 0.$$

This means that the solution u is constant on the characteristic lines of b. Hence, if we couple the transport equation (1.13) with an initial data $u(0, \cdot) = \bar{u}$, we expect

$$u(t, x) = \bar{u}\big(X(t, \cdot)^{-1}(x)\big) = \bar{u}\big(X(0, t, x)\big)$$

to be a solution of the Cauchy problem. This is easily checked with a direct computation, observing that the flow $X(t, s, x)$ satisfies the equation

$$\frac{\partial X}{\partial s}(t, s, x) + \big(b(s, x) \cdot \nabla_x\big) X(t, s, x) = 0. \qquad (1.15)$$

This is also the unique solution with initial data $\bar{u}$, since we already know that every solution has to be constant along characteristics. We then conclude the following proposition.

Proposition 1.4.2. *If the vector field b and the initial data $\bar{u}$ are C^1, then the transport equation* (1.13) *has the unique solution*

$$u(t, x) = \bar{u}\big(X(t, \cdot)^{-1}(x)\big).$$

Similarly, in the case of a transport equation with a source term $f \in C^1$ on the right hand side

$$\partial_t u(t, x) + b(t, x) \cdot \nabla u(t, x) = f(t, x),$$

we have the explicit expression

$$u(t, x) = \bar{u}(X(0, t, x)) + \int_0^t f(s, X(s, t, x))\, ds. \tag{1.16}$$

We have a similar result for the continuity equation (1.14), coupled with the initial data $\mu_0 = \bar{\mu} \in \mathcal{M}(\mathbb{R}^d)$.

Proposition 1.4.3. *Assume that the vector field b is C^1. Then, for any initial data $\bar{\mu}$, the solution of the continuity equation* (1.14) *is given by*

$$\mu_t = X(t, \cdot)_{\#}\bar{\mu}, \ \ i.e. \ \int_{\mathbb{R}^d} \varphi\, d\mu_t = \int_{\mathbb{R}^d} \varphi(X(t, x))\, d\bar{\mu}(x) \quad \forall \varphi \in C_c(\mathbb{R}^d),$$

(1.17)

where $X(t, \cdot)_{\#}\bar{\mu}$ denotes the push-forward of the measure $\bar{\mu}$ via the map $X(t, \cdot) : \mathbb{R}^d \to \mathbb{R}^d$, defined according to (A.1).

Proof. Notice first that we need only to check the distributional identity $\partial_t \mu + \text{div}\,(b\mu) = 0$ on test functions of the form $\psi(t)\varphi(x)$, that is

$$\int_{\mathbb{R}} \psi'(t)\langle \mu_t, \varphi \rangle\, dt + \int_{\mathbb{R}} \psi(t) \int_{\mathbb{R}^d} b(t, x) \cdot \nabla\varphi(x)\, d\mu_t(x)\, dt = 0. \tag{1.18}$$

We notice that the map

$$t \mapsto \langle \mu_t, \varphi \rangle = \int_{\mathbb{R}^d} \varphi(X(t, x))\, d\bar{\mu}(x)$$

belongs to $C^1([0, T])$, since the flow X is C^1 with respect to the time variable. In order to check (1.18) we need to show that the distributional derivative of this map is $\int_{\mathbb{R}^d} b(t, x) \cdot \nabla\varphi(x) \, d\mu_t(x)$, but by the C^1 regularity we only need to compute the pointwise derivative. Since the flow satisfies

$$\frac{\partial X}{\partial t}(t, x) = b(t, X(t, x))$$

for every t and x we can deduce

$$\frac{d}{dt}\langle \mu_t, \varphi \rangle = \frac{d}{dt}\int_{\mathbb{R}^d} \varphi(X(t, x)) \, d\bar{\mu}(x)$$
$$= \int_{\mathbb{R}^d} \nabla\varphi(X(t, x)) \cdot b(t, X(t, x)) \, d\bar{\mu}(x)$$
$$= \langle b(t, \cdot)\mu_t, \nabla\varphi \rangle ,$$

hence we have shown the desired thesis. $\qquad\square$

When the initial data is a locally summable function, i.e. $\bar{\mu} = \bar{\rho}\mathscr{L}^d$, we can give an explicit expression for the density ρ_t of μ_t:

$$\rho_t(x) = \frac{\bar{\rho}}{\det(\nabla X(t, \cdot))} \circ \left(X(t, \cdot)^{-1}\right)(x) .$$

This is a consequence of (1.17) and of the area formula; for a derivation of this equality see [13, Section 3].

Part I

The Eulerian
viewpoint

Chapter 2
Renormalized solutions and well-posedness of the PDE

In this chapter we begin to investigate the well-posedness of the transport equation out of the smooth setting. In the first section we present some formal computations in order to motivate the definition of *renormalized solution*, which will be central in the theory. In Section 2.2 we introduce the weak formulation of the transport equation, which is needed in order to consider the non-smooth case; it is also shown a very general result of existence of bounded solutions. In Sections 2.3 and 2.4 we introduce the notion of renormalized solution and we enlighten its importance in the well-posedness theory; we also propose a general strategy, due to DiPerna and Lions [84], which is one of the main roads to the proof of the renormalization property and thus of the well-posedness. In Sections 2.5 and 2.6 we illustrate the proof of the renormalization property in two important situations, namely the Sobolev case (DiPerna and Lions, [84]) and the BV case (Ambrosio, [8]). In the last two sections we finally present some other renormalization results due to various authors and the case of nearly incompressible vector fields.

2.1. A strategy to obtain uniqueness

In this section we describe informally one of the main tools that can be used to show the well-posedness of the transport equation

$$\begin{cases} \partial_t u(t, x) + b(t, x) \cdot \nabla u(t, x) = 0 \\ u(0, x) = \bar{u}(x) . \end{cases} \tag{2.1}$$

In order to illustrate the motivation for the concept of renormalized solution we present some computations. We proceed in a merely formal way: the argument cannot be justified without regularity assumptions on the vector field b, the uniqueness result being false in this general context (see the counterexample in Section 5.1 and the more recent ones of [5]). However, this presentation has the advantage of making evident the main points of the subsequent analysis.

We start from (2.1) and we multiply it by $2u$, obtaining

$$2u\partial_t u + 2ub \cdot \nabla u = 0.$$

Then we rewrite this equality as

$$\partial_t u^2 + b \cdot \nabla u^2 = 0. \tag{2.2}$$

Now we assume that div $b = 0$ and integrate over all $\mathbb{R}^d$, for every $t \in [0, T]$ fixed. This gives

$$\frac{d}{dt} \int_{\mathbb{R}^d} u(t, x)^2 \, dx = - \int_{\mathbb{R}^d} \text{div} \left(b(t, x) u(t, x)^2\right) dx = 0,$$

applying the divergence theorem (we also assume a sufficiently fast decay at infinity). This implies that the $L^2(\mathbb{R}^d)$ norm is conserved:

$$\frac{d}{dt} \|u(t, \cdot)\|_{L^2(\mathbb{R}^d)} = 0. \tag{2.3}$$

But the transport equation (2.1) is linear: then, in order to show uniqueness, it is enough to show that if the initial data is $\bar{u} \equiv 0$, then the only solution is $u \equiv 0$. But this is clearly implied by the conservation of the $L^2(\mathbb{R}^d)$ norm.

This formal argument has in fact two gaps. First, the application of the chain-rule to obtain (2.2) is not justified: indeed, solutions of (2.1) are not smooth in general, hence we cannot simply write

$$\partial_t u^2 = 2u\partial_t u \qquad \text{and} \qquad \nabla u^2 = 2u\nabla u.$$

The second gap is a bit more hidden, but it is also extremely relevant. The initial data in (2.1) is meant in the sense of distributions (see (2.4)). However, in order to apply (2.3) to show that $\|u(t, \cdot)\|_{L^2(\mathbb{R}^d)} = 0$ for every t, we need to know a priori that

$$\|u(t, \cdot)\|_{L^2(\mathbb{R}^d)} \to \|\bar{u}\|_{L^2(\mathbb{R}^d)} \qquad \text{as } t \downarrow 0,$$

but this is a property of strong continuity of the solution that cannot be deduced only from the weak formulation: see again the example presented in Section 5.1.

2.2. Weak solutions and existence

We first introduce the weak formulation of the transport equation (2.1). Let $b : [0, T] \times \mathbb{R}^d \to \mathbb{R}^d$ be a vector field and denote by div b the divergence of b (with respect to the spatial coordinates) in the sense of distributions.

Definition 2.2.1. If b and $\bar{u}$ are locally summable functions such that the distributional divergence of b is locally summable, then we say that a locally bounded function $u : [0, T] \times \mathbb{R}^d \to \mathbb{R}$ is a (weak) solution of (2.1) if the following identity holds for every function $\varphi \in C_c^\infty([0, T[\times\mathbb{R}^d)$:

$$\int_0^T \int_{\mathbb{R}^d} u\big[\partial_t\varphi + \varphi\operatorname{div} b + b \cdot \nabla\varphi\big]\, dx dt = -\int_{\mathbb{R}^d} \bar{u}(x)\varphi(0, x)\, dx. \quad (2.4)$$

Notice that this is the standard notion of weak solution of a PDE. For smooth solutions, equation (2.4) can be immediately deduced from (2.1) multiplying it by φ and integrating by parts.

We can equivalently define weak solutions noticing that, if $u \in L^\infty_{\mathrm{loc}}([0, T] \times \mathbb{R}^d)$, then $\partial_t u$ has a meaning as a distribution. Moreover, assuming that $\operatorname{div} b \in L^1_{\mathrm{loc}}([0, T] \times \mathbb{R}^d)$, we can define the product $b \cdot \nabla u$ as a distribution via the equality

$$\langle b \cdot \nabla u, \varphi \rangle = -\langle bu, \nabla\varphi \rangle - \langle u\operatorname{div} b, \varphi \rangle \quad \text{for every } \varphi \in C_c^\infty(]0, T[\times\mathbb{R}^d).$$

This enables us to give directly a distributional meaning to the transport equation (2.1). The initial data of the Cauchy problem (2.1) can be recovered using the following observation.

Remark 2.2.2 (Weak continuity in time). If we test the transport equation (2.1) against test functions of the form $\psi(t)\varphi(x)$ with $\psi \in C_c^1(]0, T[)$ and $\varphi \in C_c^1(\mathbb{R}^d)$ we easily obtain that

$$\frac{d}{dt} \int_{\mathbb{R}^d} u(t, x)\varphi(x)\, dx = \int_{\mathbb{R}^d} u(t, x)\big[\varphi(x)\operatorname{div} b(t, x) + b(t, x) \cdot \nabla\varphi(x)\big]\, dx$$

in the sense of distributions over $[0, T]$. This implies in particular the estimate

$$\left|\frac{d}{dt} \int_{\mathbb{R}^d} u(t, x)\varphi(x)\, dx\right| \leq \|\varphi\|_{C^1(\mathbb{R}^d)} V_R(t) \quad (2.5)$$

for every φ with $\operatorname{spt}\varphi \subset B_R(0)$, where

$$V_R(t) = \|u\|_\infty \int_{B_R(0)} \big(|\operatorname{div} b(t, x)| + |b(t, x)|\big)\, dx \in L^1([0, T]).$$

Let $L_\varphi \subset [0, T]$ be the set of the Lebesgue points of the map $t \mapsto \int u(t, x)\varphi(x)\, dx$. We know that $\mathcal{L}^1([0, T] \setminus L_\varphi) = 0$. Now consider for every $R \in \mathbb{N}$ a countable set $\mathcal{Z}_R$ which is dense in $C_c^1(B_R(0))$ with respect to the C^1 norm and set $L_{\mathcal{Z}_R} = \cap_{\varphi \in \mathcal{Z}_R} L_\varphi$. Clearly we have

$\mathscr{L}^1([0, T] \setminus L_{\mathcal{Z}_R}) = 0$. The restriction of $u(t, \cdot)$ to $L_{\mathcal{Z}_R}$ provides a uniformly continuous family of bounded functionals on $C_c^1(B_R(0))$, since estimate (2.5) implies

$$\left| \int_{\mathbb{R}^d} u(t, x)\varphi(x)\,dx - \int_{\mathbb{R}^d} u(s, x)\varphi(x)\,dx \right| \le \|\varphi\|_{C^1} \int_s^t V_R(\tau)\,d\tau$$

for every $\varphi \in C_c^1(B_R(0))$ and every $s, t \in L_{\mathcal{Z}_R}$. Therefore $u(t, \cdot)$ can be extended in a unique way to a continuous curve $\tilde{u}_R(t, \cdot)$ in $\left[C_c^1(B_R(0)) \right]'$, for $t \in [0, T]$. Applying iteratively this argument a countable number of times, we can construct in a unique way a continuous curve $\tilde{u}(t, \cdot)$ in $\left[C_c^1(B_R(0)) \right]'$, for $t \in [0, T]$ and for every $R \in \mathbb{N}$. Recalling that $u(t, x) \in L^\infty([0, T] \times \mathbb{R}^d)$ this also implies that $\tilde{u}(t, \cdot)$ is a continuous curve in $\left[L^1(\mathbb{R}^d) \right]'$.

From the above argument, which is indeed classical in the theory of evolutionary PDEs (see for instance [18, Lemma 8.1.2] or [75, Theorem 4.1.1]), we deduce that, up to a modification of $u(t, \cdot)$ in a negligible set of times, we can assume that $t \mapsto u(t, \cdot)$ is weakly* continuous from $[0, T]$ into $L^\infty(\mathbb{R}^d)$. A similar remark applies to the continuity equation (1.14): we can assume that the map $t \mapsto \mu_t$ is weakly* continuous from $[0, T]$ to $\mathcal{M}(\mathbb{R}^d)$. It follows in particular that $u(t, \cdot)$ and μ_t are defined for *every* $t \in [0, T]$, and in particular at the endpoints; this also gives a sense to the Cauchy data at $t = 0$. As it will be clear from the example of Section 5.1, in general we cannot expect *strong* continuity of the solution with respect to the time.

Existence of weak solutions is quite a trivial issue: since the transport equation is a linear PDE, it is sufficient to regularize the vector field and the initial data, obtaining a sequence of smooth solutions to the approximate problems, and then we pass to the limit to get a solution.

Theorem 2.2.3 (Existence of weak solutions). *Let* $b \in L^\infty([0, T] \times \mathbb{R}^d; \mathbb{R}^d)$ *with* $\text{div}\, b \in L_{\text{loc}}^1([0, T] \times \mathbb{R}^d)$ *and let* $\bar{u} \in L^\infty(\mathbb{R}^d)$. *Then there exists a weak solution* $u \in L^\infty([0, T] \times \mathbb{R}^d)$ *to (2.1).*

Proof. Let ρ_ϵ be a convolution kernel on $\mathbb{R}^d$ and η_ϵ be a convolution kernel on $\mathbb{R}^{1+d}$. We define $\bar{u}^\epsilon = \bar{u} * \rho_\epsilon$ and $b^\epsilon = b * \eta_\epsilon$. Let u^ϵ be the unique solution to the regularized transport equation

$$\begin{cases} \partial_t u + b^\epsilon \cdot \nabla u = 0 \\ u(0, \cdot) = \bar{u}^\epsilon. \end{cases}$$

This solution exists and is unique for every $\epsilon > 0$, since the vector field b^ϵ is smooth. From the explicit formula (1.16), for the solution of the transport equation with smooth vector field, we immediately deduce that $\{u^\epsilon\}$ is equi-bounded in $L^\infty([0, T] \times \mathbb{R}^d)$. Thus we can find a subsequence $\{u^{\epsilon_j}\}_j$ which converges in $L^\infty([0, T] \times \mathbb{R}^d) - w^*$ to a function $u \in L^\infty([0, T] \times \mathbb{R}^d)$. Recalling the weak formulation (2.4) it is immediate to deduce that u is a solution of (2.1). $\square$

2.3. Renormalized solutions

We now want to face the uniqueness issue. As we already noticed, the application of the chain-rule that we needed in the computations in Section 2.1 is not always justified. We define here the notion of renormalized solution, which corresponds to the idea of "solution which satisfies the desired chain-rule".

Definition 2.3.1 (Renormalized solution). Let $b : I \times \mathbb{R}^d \to \mathbb{R}^d$ be a locally summable vector field such that div b is locally summable, where $I \subset \mathbb{R}$ is an interval. Let $u \in L^\infty(I \times \mathbb{R}^d)$ be a solution of the transport equation

$$\partial_t u + b \cdot \nabla u = 0.$$

We say that u is a *renormalized solution* if the equation

$$\partial_t \beta(u) + b \cdot \nabla \beta(u) = 0 \tag{2.6}$$

holds in the sense of distributions in $I \times \mathbb{R}^d$ for every function $\beta \in C^1(\mathbb{R}; \mathbb{R})$.

This property, if satisfied by all bounded weak solutions, can be transferred into a property of the vector field itself.

Definition 2.3.2 (Renormalization property). Let $b : I \times \mathbb{R}^d \to \mathbb{R}^d$ be a locally summable vector field such that div b is locally summable, where $I \subset \mathbb{R}$ is an interval. We say that b has the *renormalization property* if every bounded solution of the transport equation with vector field b is a renormalized solution.

The importance of the renormalization property is summarized in the following theorem, which corresponds to the rough statement "renormalization implies well-posedness".

Theorem 2.3.3. *Let $b : [0, T] \times \mathbb{R}^d \to \mathbb{R}^d$ be a bounded vector field with* div $b \in L^1([0, T]; L^\infty(\mathbb{R}^d))$. *Define the vector field $\tilde{b} :]-\infty, T] \times \mathbb{R}^d \to \mathbb{R}^d$ according to*

$$\tilde{b}(t, x) = \begin{cases} 0 & \text{if } t < 0 \\ b(t, x) & \text{if } 0 \leq t \leq T. \end{cases} \tag{2.7}$$

If $\tilde{b}$ has the renormalization property, then bounded solutions of the transport equation (2.1) are unique *and* stable. *By stability we mean the following: let b_k and $\bar{u}_k$ be two smooth approximating sequences converging strongly in L^1_{loc} to b and $\bar{u}$ respectively, with $\|\bar{u}_k\|_\infty$ uniformly bounded; then the solutions u_k of the corresponding transport equations converge strongly in L^1_{loc} to the solution u of (2.1).*

Proof. UNIQUENESS. Since the transport equation is linear, it is sufficient to show that the only bounded solution to the Cauchy problem

$$\begin{cases} \partial_t u + b \cdot \nabla u = 0 \\ u(0, \cdot) = 0 \end{cases} \qquad \text{in } \mathcal{D}'([0, T] \times \mathbb{R}^d)$$

is $u \equiv 0$. Extending b to negative times as in the statement of the theorem, we obtain that the following equation is satisfied, for $t \in] - \infty, T]$:

$$\begin{cases} \partial_t u + \tilde{b} \cdot \nabla u = 0 \\ u(t, \cdot) = 0 \quad \text{for } t \le 0 \end{cases} \qquad \text{in } \mathcal{D}'(] - \infty, T] \times \mathbb{R}^d).$$

Since we are assuming that $\tilde{b}$ has the renormalization property we deduce that

$$\begin{cases} \partial_t u^2 + \tilde{b} \cdot \nabla u^2 = 0 \\ u(t, \cdot) = 0 \quad \text{for } t \le 0 \end{cases} \qquad \text{in } \mathcal{D}'(] - \infty, T] \times \mathbb{R}^d). \qquad (2.8)$$

Now fix $R > 0$ and $\eta > 0$ and consider a test function $\varphi \in C_c^\infty(] - \infty, T[\times \mathbb{R}^d)$ such that $\varphi = 1$ on $[0, T - \eta] \times B_R(0)$ and

$$\partial_t \varphi(t, x) \le -\|b\|_\infty |\nabla \varphi(t, x)| \qquad \text{on } [0, T] \times \mathbb{R}^d. \qquad (2.9)$$

We define

$$f(t) = \int_{\mathbb{R}^d} u^2(t, x) \varphi(t, x) \, dx \, .$$

For every positive $\psi \in C_c^\infty(] - \infty, T[)$ we test (2.8) against $\psi(t) \varphi(t, x)$ and we apply Fubini's theorem to deduce

$$- \int_{-\infty}^T f(t) \psi'(t) \, dt$$

$$= \int_{-\infty}^T \int_{\mathbb{R}^d} u^2(t, x) \psi(t) \left[\partial_t \varphi(t, x) + \tilde{b}(t, x) \cdot \nabla \varphi(t, x) \right] dx \, dt$$

$$+ \int_{-\infty}^T f(t) \operatorname{div} \tilde{b}(t, x) \psi(t) \, dt$$

$$\overset{(2.9)}{\le} \int_{-\infty}^T f(t) \operatorname{div} \tilde{b}(t, x) \psi(t) \, dt$$

$$\le \int_{-\infty}^T f(t) \|\operatorname{div} \tilde{b}(t, \cdot)\|_{L^\infty(\mathbb{R}^d)} \psi(t) \, dt \, .$$

This means that we have

$$\frac{d}{dt} f(t) \leq \|\operatorname{div} \tilde{b}(t, \cdot)\|_{L^{\infty}(\mathbb{R}^d)} f(t)$$

in the sense of distributions in $] - \infty, T]$. But we know, by (2.8), that $f(t) = 0$ for $t < 0$. Using the assumption $\operatorname{div} b \in L^1([0, T]; L^{\infty}(\mathbb{R}^d))$ (which clearly implies $\operatorname{div} \tilde{b} \in L^1(] - \infty, T]; L^{\infty}(\mathbb{R}^d)))$ we can apply the Gronwall lemma and deduce that

$$f(t) = \int_{\mathbb{R}^d} u^2(t, x) \varphi(t, x) \, dx = 0$$

for every $t \in] - \infty, T]$. From the arbitrariness of $\varphi(t, x)$ we deduce that $u(t, \cdot) = 0$ for every $t \in [0, T]$, as desired.

STABILITY. Arguing as in Theorem 2.2.3, we easily deduce that, up to subsequences, u_k converges in $L^{\infty}([0, T] \times \mathbb{R}^d) - w^*$ to a distributional solution u of (2.1). However, by the uniqueness part of the theorem, we know that this solution is unique, and then the whole sequence converges to u. Since b_k and u_k are both smooth, u_k^2 solves the transport equation with vector field b_k and initial data $\bar{u}_k^2$. Arguing as before, u_k^2 must converge in $L^{\infty}([0, T] \times \mathbb{R}^d) - w^*$ to the unique solution of (2.1) with initial data $\bar{u}^2$. But by the renormalization property this solution is u^2. Hence we have shown that $u_k \overset{*}{\rightharpoonup} u$ and $u_k^2 \overset{*}{\rightharpoonup} u^2$ in $L^{\infty}([0, T] \times \mathbb{R}^d) - w^*$, and this eventually implies $u_k \to u$ strongly in $L^1_{\text{loc}}([0, T] \times \mathbb{R}^d)$. □

Remark 2.3.4. In the previous theorem the extension to negative times in (2.7) is necessary: it is a way of requiring the strong continuity of the solution at the initial time (see the discussion in Section 2.1 and Theorem 3.1.1). Another possibility would be to change the definition of renormalized solution, asking that $\beta(u)$ should satisfy the transport equation and have initial data $\beta(\bar{u})$ (this is the approach of [77] and [79]).

The renormalization property for $\tilde{b}$ in general does not follow from the renormalization property for b. Moreover, even if b has the renormalization property in the sense of Definition 2.3.2, it could exist a solution u which is renormalized, but such that $\beta(u)(0, \cdot) \neq \beta(\bar{u})$. An explicit example is given by the oscillatory solution constructed by Depauw [82] and presented in Section 5.1. In this case the vector field b has the renormalization property (in $[0, 1] \times \mathbb{R}^2$) but the extension $\tilde{b}$ has not the renormalization property in $] - \infty, 1] \times \mathbb{R}^2$; the Cauchy problem for the transport equation with initial data $\bar{u} = 0$ has more than one solution. This example also shows the sharpness of the assumptions of Theorem 2.6.1: indeed, $b \in L^1_{\text{loc}}(]0, 1[; BV_{\text{loc}}(\mathbb{R}^2; \mathbb{R}^2))$ but $b \notin L^1_{\text{loc}}([0, 1[; BV_{\text{loc}}(\mathbb{R}^2; \mathbb{R}^2))$, thus the extension $\tilde{b} \notin L^1_{\text{loc}}(] - \infty, 1[; BV_{\text{loc}}(\mathbb{R}^2; \mathbb{R}^2))$.

Remark 2.3.5. The result in Theorem 2.3.3, together with the renormalization results we are going to prove in the rest of this chapter, extends to transport equations with a linear right hand side of order zero in u of the form

$$\partial_t u + b \cdot \nabla u = cu ,$$

with $c \in L^1([0, T]; L^\infty_{\mathrm{loc}}(\mathbb{R}^d))$. The only modification in the proof consists in taking into account an additional term in the Gronwall argument. In particular, choosing $c = -\operatorname{div} b$, we are able to translate all the forthcoming well-posedness results for the transport equation into well-posedeness results for the continuity equation.

Remark 2.3.6.   When dealing with the continuity equation $\partial_t u + \operatorname{div}(bu) = 0$ we can consider the renormalization function $\beta(z) = \sqrt{1 + (z^+)^2} - 1 \in C^1(\mathbb{R})$. Integrating formally over $\mathbb{R}^d$ we obtain

$$\frac{d}{dt} \int_{\mathbb{R}^d} \beta(u(t, x)) dx = \int_{\mathbb{R}^d} \operatorname{div} b(t, x) \big[\beta(u(t, x)) - u(t, x)\beta'(u(t, x))\big] dx .$$

Since for this choice of β we have $\beta(z) - z\beta'(z) \leq 0$, it turns out that it is enough to have a control on the negative part of the divergence: we only need to assume $[\operatorname{div} b]^- \in L^1([0, T]; L^\infty(\mathbb{R}^d))$.

Since in all the forthcoming renormalization results no regularity (but just integrability) is required with respect to the time, it is clear that the zero extension for negative times remains in the same class of regularity initially assumed for the vector field. We will deal with vector fields defined on $I \times \mathbb{R}^d$, where $I \subset \mathbb{R}$ is a generic interval of times, since in order to treat the well-posedness problem for $t \in [0, T]$ we need to show the renormalization property for the extended vector field, defined on the time interval $I =] -\infty, T]$. In the following we are going to illustrate various cases in which the renormalization property holds: in all these cases the well-posedness for the transport equation and for the continuity equation presented in Theorem 2.3.3 will then hold.

2.4. The DiPerna–Lions regularization scheme

As we remarked in the initial discussion of Section 2.1, the main obstruction to the renormalization property is the lack of regularity of the solution u. It is therefore not surprising that the main technique used to prove this property is a regularization procedure. This idea goes back to the work by DiPerna and Lions [84], where it was used to show the renormalization property for Sobolev vector fields. The link between renormalization and more general regularization procedures will be also investigated in the next chapter: see in particular Theorem 3.1.1 and Remark 3.1.6.

Let us fix an even convolution kernel ρ_ϵ in $\mathbb{R}^d$. We start from the transport equation $\partial_t u + b \cdot \nabla u = 0$ and we convolve it by ρ_ϵ. We define $u^\epsilon = u * \rho_\epsilon$ and we notice that we can let the convolution act on u in $b \cdot \nabla u$ only at the price of an error term:

$$\partial_t u^\epsilon + b \cdot \nabla u^\epsilon = b \cdot \nabla u^\epsilon - (b \cdot \nabla u) * \rho_\epsilon . \qquad (2.10)$$

We define the *commutator* r^ϵ as the error term in the right hand side of the previous equality:

$$r^\epsilon = b \cdot \nabla u^\epsilon - (b \cdot \nabla u) * \rho_\epsilon . \qquad (2.11)$$

We call this term commutator because it measures the difference in exchanging the two operations of convolution and of differentiation in the direction of b. Now notice that the function u^ϵ is smooth with respect to the spatial variable. Moreover we have the identity

$$\partial_t u^\epsilon = -(b \cdot \nabla u) * \rho_\epsilon ,$$

hence, for every $\epsilon > 0$ fixed, the function u^ϵ has Sobolev regularity in space-time. This implies that, if we multiply both sides of (2.10) by $\beta'(u^\epsilon)$, we can apply Stampacchia's chain-rule for Sobolev maps (see for instance [91, Section 4.2.2]) to get

$$\beta'(u^\epsilon)\big[\partial_t u^\epsilon + b \cdot \nabla u^\epsilon\big] = \partial_t \beta(u^\epsilon) + b \cdot \nabla \beta(u^\epsilon) .$$

Thus, for every $\epsilon > 0$ we have the equality

$$\partial_t \beta(u^\epsilon) + b \cdot \nabla \beta(u^\epsilon) = r^\epsilon \beta'(u^\epsilon) . \qquad (2.12)$$

Now we would like to pass to the limit as $\epsilon \to 0$ in order to recover the renormalization property. The left hand side converges in the sense of distributions to the left hand side of (2.6). Thus we need to show the convergence to zero of the quantity $r^\epsilon \beta'(u^\epsilon)$.

Notice that r^ϵ converges to zero in the sense of distributions, without any regularity assumption on b. The sequence $\beta'(u^\epsilon)$ is equi-bounded in L^∞ and converges $\mathscr{L}^d$-a.e. for every $t \in I$, but this does not allow to deduce the convergence of the product $r^\epsilon \beta'(u^\epsilon)$. A first attempt would be to show the strong convergence to zero of the commutator r^ϵ. This is precisely the result that DiPerna and Lions obtained in [84] under a Sobolev regularity assumption.

2.5. Vector fields with Sobolev regularity

Theorem 2.5.1 (DiPerna–Lions). *Let b be a bounded vector field belonging to $L^1_{loc}(I; W^{1,p}_{loc}(\mathbb{R}^d; \mathbb{R}^d))$, where $I \subset \mathbb{R}$ is an interval. Then b has the renormalization property.*

Proof. The theorem immediately follows from the discussion in the previous section and from the following proposition, in which strong convergence to zero of the commutator is proved. □

Proposition 2.5.2 (Strong convergence of the commutator). *Let b be a bounded vector field belonging to $L^1_{loc}(I; W^{1,p}_{loc}(\mathbb{R}^d; \mathbb{R}^d))$, where $I \subset \mathbb{R}$ is an interval, and let $u \in L^\infty_{loc}(I \times \mathbb{R}^d)$. Define the commutator r^ϵ as in (2.11). Then $r^\epsilon \to 0$ strongly in $L^1_{loc}(I \times \mathbb{R}^d)$.*

Proof. Playing with the definitions of $b \cdot \nabla u$ and of the convolution of a distribution and a smooth function it is not difficult to show the following explicit expression of the commutator:

$$r^\epsilon(t, x) = \int_{\mathbb{R}^d} u(t, x - \epsilon z) \frac{b(t, x) - b(t, x - \epsilon z)}{\epsilon} \cdot \nabla \rho(z) \, dz + \big(u \operatorname{div} b\big) * \rho_\epsilon \,.$$

$$(2.13)$$

We recall the following property, which in fact characterizes functions in Sobolev spaces (see [131, Theorem 2.1.6]): for every $f \in W^{1,1}_{loc}$ we have

$$\frac{f(x + \epsilon z) - f(x)}{\epsilon} \to \nabla f(x) z \qquad \text{strongly in } L^1_{loc} \text{ as } \epsilon \to 0, \quad (2.14)$$

that is, the difference quotients converge strongly to the derivative. Using (2.14) and the strong convergence of translations in L^p in (2.13) we obtain that r^ϵ converges strongly in $L^1_{loc}(I \times \mathbb{R}^d)$ to

$$u(t, x) \int_{\mathbb{R}^d} \big(\nabla b(t, x) z\big) \cdot \nabla \rho(z) \, dz + u \operatorname{div} b \,.$$

The elementary identity

$$\int_{\mathbb{R}^d} z_i \frac{\partial \rho(z)}{\partial z_j} \, dz = -\delta_{ij}$$

immediately shows that the limit is zero. This concludes the proof. □

2.6. Vector fields with bounded variation

In this section we present the main lines of the argument by Ambrosio [8] for the renormalization for vector fields with bounded variation. The main point which makes the difference with respect to the Sobolev case is the lack of strong convergence of the difference quotients (compare with (2.14)). The proof will involve a splitting of the difference quotients and an anisotropic regularization, based on a local selection of a "bad" direction given by Alberti's Rank-one theorem.

Theorem 2.6.1 (Ambrosio). *Let b be a bounded vector field belonging to $L^1_{\mathrm{loc}}(I; BV_{\mathrm{loc}}(\mathbb{R}^d; \mathbb{R}^d))$, such that $\mathrm{div}\, b \ll \mathscr{L}^d$ for $\mathscr{L}^1$-a.e. $t \in I$. Then b has the renormalization property.*

We intend to give a reasonably detailed description of the proof of this result, mostly trying to enlighten the main ideas. We will present an approach which is slightly simpler than the one of the original paper [8]. We also refer to [9, 10, 79] for an account of the proof of Ambrosio's theorem. This theorem has been applied to the study of various nonlinear PDEs: for instance the Keyfitz and Kranzer system ([15, 11]) and the semigeostrophic equation ([72, 73, 31]).

We start by introducing some notation. For the basic terminology and facts about BV functions we refer to Appendix A.3. For every $t \in I$ we perform the decomposition of the spatial derivative of $b(t, \cdot)$ (which is, by definition, a matrix-valued measure in $\mathbb{R}^d$) in absolutely continuous and singular part as follows:

$$Db(t, \cdot) = D^a b(t, \cdot) + D^s b(t, \cdot).$$

Moreover, we denote by $|Db|$, $|D^a b|$ and $|D^s b|$ the measures obtained by integration of $|Db(t, \cdot)|$, $|D^a b(t, \cdot)|$ and $|D^s b(t, \cdot)|$ with respect to the time variable, that is

$$\int_{I \times \mathbb{R}^d} \varphi(t, x) d|Db|(t, x) = \int_I \int_{\mathbb{R}^d} \varphi(t, x) d|Db(t, \cdot)|(x)\, dt,$$

$$\int_{I \times \mathbb{R}^d} \varphi(t, x) d|D^\sigma b|(t, x) = \int_I \int_{\mathbb{R}^d} \varphi(t, x) d|D^\sigma b(t, \cdot)|(x)\, dt, \quad \sigma = a, s,$$

for every function $\varphi \in C_c(I \times \mathbb{R}^d)$.

2.6.1. Difference quotients of BV functions

As we already observed, the strong convergence of the difference quotients in (2.14) characterizes functions in Sobolev spaces. Hence a first step in the proof of Theorem 2.6.1 is a description of the behaviour of the difference quotients for a function with bounded variation.

For any function $f \in BV_{\mathrm{loc}}$ and any $z \in \mathbb{R}^d$ with $|z| \le \epsilon$ we have the classical L^1 estimate of the increments

$$\int_K |f(x+z) - f(x)|\, dx \le |D_z f|(K_\epsilon) \qquad \text{for any compact set } K \subset \mathbb{R}^d,$$

$$(2.15)$$

where $D_z f = (Df)z$ denotes the component of Df in the direction z and K_ϵ is the open ϵ-neighbourhood of K.

Moreover, we can give a canonical decomposition of the difference quotients in order to analyse independently the behaviour of the absolutely continuous part and of the singular part. Let us start from the case $d = 1$. If μ is an $\mathbb{R}^m$-valued measure in $\mathbb{R}$ with locally finite variation, then by Jensen's inequality the functions

$$\widehat{\mu}_\epsilon(t) = \frac{\mu([t, t+\epsilon])}{\epsilon}; \qquad t \in \mathbb{R}$$

satisfy

$$\int_K |\widehat{\mu}_\epsilon(t)|\, dt \le |\mu|(K_\epsilon) \qquad \text{for every compact set } K \subset \mathbb{R}.$$

A simple density argument then shows that $\widehat{\mu}_\epsilon$ converge in $L^1_{\mathrm{loc}}(\mathbb{R})$ to the density of μ with respect to $\mathscr{L}^1$ whenever $\mu \ll \mathscr{L}^1$. If $f \in BV_{\mathrm{loc}}(\mathbb{R}; \mathbb{R}^m)$ and $\epsilon > 0$ we know that

$$\frac{f(x+\epsilon) - f(x)}{\epsilon} = \frac{Df([x, x+\epsilon])}{\epsilon}$$

$$= \frac{D^a f([x, x+\epsilon])}{\epsilon} + \frac{D^s f([x, x+\epsilon])}{\epsilon}$$

for $\mathscr{L}^1$-a.e. $x \in \mathbb{R}$, the exceptional set possibly depending on ϵ. In this way we have canonically split the difference quotient of f as the sum of two functions, one strongly convergent to the absolutely continuous part of the derivative $D^a f$, and the other one having an L^1 norm on any compact set K asymptotically smaller than $|D^s f|(K)$.

This argument can be carried on also in the multi-dimensional case with the help of the slicing theory of BV functions. If we fix the direction $z \in \mathbb{R}^d$, we obtain that the difference quotients can be canonically split into two parts as follows

$$\frac{b(t, x) - b(t, x - \epsilon z)}{\epsilon} = b^1_{\epsilon, z}(t, x) + b^2_{\epsilon, z}(t, x), \qquad (2.16)$$

with

$$b^1_{\epsilon,z}(t,x) \to D^a b(t,x)\, z \qquad \text{strongly in } L^1_{\text{loc}}(\mathbb{R}^d) \qquad (2.17)$$

and

$$\limsup_{\epsilon \to 0} \int_K |b^2_{\epsilon,z}(t,x)|\, dx \le |D^s b(t,x)z|(K) \text{ for every compact set } K \subset \mathbb{R}^d.$$
$$(2.18)$$

2.6.2. Isotropic estimate and the defect measure σ

We proceed with the DiPerna–Lions regularization scheme illustrated in Section 2.4. We arrive at the expression (2.13) for the commutator and we insert in it the decomposition (2.16) of the difference quotients, getting

$$r^\epsilon = b \cdot \nabla u^\epsilon - (b \cdot \nabla u) * \rho_\epsilon$$
$$= \int_{\mathbb{R}^d} u(t, x - \epsilon z)\frac{b(t,x) - b(t, x - \epsilon z)}{\epsilon} \cdot \nabla \rho(z)\, dz + (u \operatorname{div} b) * \rho_\epsilon$$
$$= \int_{\mathbb{R}^d} u(t, x - \epsilon z) b^1_{\epsilon,z}(t,x) \cdot \nabla \rho(z)\, dz + (u \operatorname{div} b) * \rho_\epsilon \qquad (2.19)$$
$$+ \int_{\mathbb{R}^d} u(t, x - \epsilon z) b^2_{\epsilon,z}(t,x) \cdot \nabla \rho(z)\, dz. \qquad (2.20)$$

Arguing as in Section 2.5, applying (2.17) we deduce that (2.19) converges strongly to

$$u(t,x) \int_{\mathbb{R}^d} D^a b(t,x) z \cdot \nabla \rho(z)\, dz + u(t,x) \operatorname{div} b(t,x)$$
$$= -u(t,x) \operatorname{trace}\big[D^a b(t,x)\big] + u(t,x) \operatorname{div} b(t,x) = 0,$$

since $\operatorname{trace}\big[D^a b(t,x)\big]$ is the absolutely continuous part of the divergence, which however coincides with the whole divergence, because we are assuming that $\operatorname{div} b \ll \mathscr{L}^d$.

For the integral in (2.20) we use the estimate (2.18) for $b^2_{\epsilon,z}$ to deduce the isotropic estimate

$$\limsup_{\epsilon \to 0} \int_J \int_K |r^\epsilon|\, dt dx \le \|u\|_\infty I(\rho)|D^s b|(J \times K), \qquad (2.21)$$

for every compact set $K \subset \mathbb{R}^d$ and every interval $J \subset I$, where we have defined the *isotropic energy* of the convolution kernel ρ to be

$$I(\rho) = \int_{\mathbb{R}^d} |z|\, |\nabla \rho(z)|\, dz.$$

From the isotropic estimate (2.21) and the equi-boundedness in $L^\infty(I \times \mathbb{R}^d)$ of $\{\beta'(u^\epsilon)\}$ we deduce that the sequence $\{r^\epsilon \beta'(u^\epsilon)\}$ has limit points in the sense of measures. Hence we can introduce the *defect measure*

$$\sigma = \partial_t \beta(u) + b \cdot \nabla \beta(u) \qquad (2.22)$$

and we deduce from (2.21) that σ is a locally finite measure satisfying

$$|\sigma| \leq \|\beta'\|_\infty \|u\|_\infty I(\rho)|D^s b| \qquad \text{as measures on } I \times \mathbb{R}^d. \qquad (2.23)$$

In particular, this shows that σ is singular with respect to $\mathscr{L}^1 \otimes \mathscr{L}^d$. We can interpret (2.22) by saying that, up to now, we have shown the renormalization property up to an error term which is a singular measure.

2.6.3. Anisotropic estimate

We now give another estimate for the local norm of the commutator r^ϵ. In this estimate we do not take any advantage of the cancellations, but we introduce an anisotropic term related to the derivative of the vector field b which will be useful in the final estimate.

We denote by M_t the matrix which appears in the polar decomposition of $Db(t, \cdot)$, that is the $(d \times d)$-matrix defined $|Db(t, \cdot)|$-a.e. by the identity $Db(t, \cdot) = M_t(\cdot)|Db(t, \cdot)|$. For every $z \in \mathbb{R}^d$ and for $\mathscr{L}^1$-a.e. $t \in I$ we have the identity

$$D_z b(t, \cdot) \cdot \nabla \rho(z) = M_t(\cdot) z \cdot \nabla \rho(z)|Db(t, \cdot)| \quad \text{as measures on } \mathbb{R}^d. \quad (2.24)$$

Going back to (2.13) and using (2.24) and the L^1 estimate (2.15) we obtain

$$\limsup_{\epsilon \to 0} \int_J \int_K |r^\epsilon| dt dx \leq \int_J \int_K \Lambda(M_t(x), \rho) d|Db|(t, x) + d|D^a b|(J \times K) \qquad (2.25)$$

for every compact set $K \subset \mathbb{R}^d$ and every interval $J \subset I$. Here, for every $d \times d$ matrix N and for every convolution kernel ρ, we have defined the *anisotropic energy* of ρ as

$$\Lambda(N, \rho) = \int_{\mathbb{R}^d} |Nz \cdot \nabla \rho(z)| \, dz.$$

This implies that the defect measure σ satisfies

$$|\sigma| \leq \|\beta'\|_\infty \|u\|_\infty \big[\Lambda(M_t(\cdot), \rho)|Db| + d|D^a b|\big] \quad \text{as measures on } I \times \mathbb{R}^d. \qquad (2.26)$$

2.6.4. Combination of the two estimates

From (2.23) and (2.26) we obtain that

$$|\sigma| \leq \|\beta'\|_\infty \|u\|_\infty \Lambda(M_t(\cdot), \rho)|D^s b| \quad \text{as measures on } I \times \mathbb{R}^d. \quad (2.27)$$

This can be deduced from the following general fact: if σ, μ_1 and μ_2 are positive measures such that $\sigma \leq \mu_1 + \mu_2$ and $\sigma \perp \mu_2$, then $\sigma \leq \mu_1$.

Now we observe that (2.27) holds for every even convolution kernel ρ. The important point here is that the defect measure σ does not depend on ρ. Our aim is to show that $\sigma = 0$, and this will be achieved optimizing this estimate. Since the estimate has a local nature, this optimization procedure is, in a certain sense, equivalent to vary the regularizing kernel in t and x. We claim that

$$|\sigma| \leq \|\beta'\|_\infty \|u\|_\infty \left[\inf_\rho \Lambda(M_t(\cdot), \rho)\right]|D^s b|, \quad (2.28)$$

where the infimum is taken over the set

$$\mathcal{K} = \left\{\rho \in \mathcal{C}_c^\infty(\mathbb{R}^d), \text{ such that } \rho \geq 0 \text{ is even and } \int_{\mathbb{R}^d} \rho = 1\right\}.$$

We first notice that (2.27) gives that $\sigma \ll |D^s b|$, hence there exists a Borel function $f(t, x)$ such that

$$\sigma = f(t, x)|D^s b|. \quad (2.29)$$

Hence we can rewrite (2.27) as

$$|f(t, x)| \leq \|\beta'\|_\infty \|u\|_\infty \Lambda(M_t(x), \rho) \quad \text{for } |D^s b|\text{-a.e. } (t, x) \quad (2.30)$$

for every fixed ρ. Notice however that the set where the inequality (2.30) fails could in principle depend on ρ. This is not a problem as soon as we infimize on a countable set of kernels $\mathcal{K}' \subset \mathcal{K}$: we can deduce

$$|f(t, x)| \leq \|\beta'\|_\infty \|u\|_\infty \left[\inf_{\rho \in \mathcal{K}'} \Lambda(M_t(x), \rho)\right] \quad \text{for } |D^s b|\text{-a.e. } (t, x). \quad (2.31)$$

However we notice that, for every fixed matrix M, the map $\rho \mapsto \Lambda(M, \rho)$ is continuous with respect to the $W^{1,1}$ topology. Choosing $\mathcal{K}' \subset \mathcal{K}$ countable and dense with respect to the $W^{1,1}$ topology we obtain that the infimum over $\mathcal{K}'$ coincides with the infimum over $\mathcal{K}$. This implies (2.28).

2.6.5. Local optimization of the convolution kernel

In order to show that $\sigma = 0$ and to obtain the renormalization property we only need to show that the infimum in (2.28) is actually equal to zero, under the assumption that b has bounded variation and div b is absolutely continuous with respect to $\mathscr{L}^d$. The original proof in [8] is based on Alberti's Rank-one theorem (see Theorem A.3.2) and on an anisotropic regularization procedure (first introduced by Bouchut [36] in the context of the Vlasov equation with BV field and subsequently exploited by Colombini and Lerner [60] to treat the transport equation with conormal-BV vector field). We present in the next subsection a simplified version of the proof, based on a lemma due to Alberti, in which the infimum in (2.28) is computed.

The anisotropic regularization procedure of Bouchut is based on the following heuristic remark. Suppose that the vector field b depends "nicely" on a group of variables x_1 and "badly" on another group x_2. Then it is more efficient to regularize quickly along the direction x_2 and slowly along the direction x_1, in order to rule out faster the bad dependence in x_2. This can be achieved using anisotropic convolution kernels, which are asymptotically very close to the characteristic function of a very thin rectangle, whose long side is along the x_1 direction.

This is applied by Colombini and Lerner [60] to deal with the case of conormal-BV vector fields, in which the derivative of b is a measure along one direction and an L^1 function along the other $d - 1$ directions. Then they regularize much faster in the direction along which the derivative is a measure, according to Bouchut's scheme.

In [8] the selection of the two directions is based on Alberti's Rank-one theorem (see Theorem A.3.2). This result allows to select, for $|D^s b(t, \cdot)|$-a.e. $x \in \mathbb{R}^d$, two unit vectors ξ_t and η_t in such a way that

$$D^s b(t, \cdot) = \xi_t \otimes \eta_t |D^s b(t, \cdot)| \,.$$

This can be intepreted as an asymptotic dependence of $b(t, \cdot)$ on the direction η_t only, hence the shape of the anisotropic kernel is carefully chosen to be very thin in this direction.

2.6.6. Alberti's lemma

In this section we describe Alberti's lemma, which allows to compute exactly the infimum

$$\Lambda(M) = \inf_{\rho \in \mathcal{K}} \Lambda(M, \rho) = \inf_{\rho \in \mathcal{K}} \int_{\mathbb{R}^d} |Mz \cdot \nabla \rho(z)| dz,$$

for every $d \times d$ matrix M. The proof is again related to an anisotropic regularization: the basic idea is in some sense to take a convolution kernel which is concentrated on a very long tube made of trajectories of the ordinary differential equation $\dot{\gamma} = M\gamma$.

Lemma 2.6.2. *For every $d \times d$ matrix M we have*

$$\Lambda(M) = |\operatorname{trace}(M)|.$$

We observe first that this result allows to conclude the proof of Theorem 2.6.1. Indeed, from the assumption $\operatorname{div} b \ll \mathscr{L}^d$ for $\mathscr{L}^1$-a.e. $t \in I$, it follows that

$$\operatorname{trace}(M_t(\cdot))|D^s b(t, \cdot)| = 0 \qquad \text{for } \mathscr{L}^1\text{-a.e. } t \in I.$$

This means that

$$\inf_{\rho \in \mathcal{K}} \Lambda(M_t(\cdot), \rho) = 0$$

for $|D^s b|$-a.e. $(t, x) \in I \times \mathbb{R}^d$. Going back to (2.28) we deduce that $\sigma = 0$, hence we have proved the renormalization result of Theorem 2.6.1.

Proof of Lemma 2.6.2. We first observe that the inequality $\Lambda(M) \geq |\operatorname{trace}(M)|$ is trivial. Indeed for every admissible kernel ρ we have

$$\int_{\mathbb{R}^d} |Mz \cdot \nabla\rho(z)| \, dz \geq \left| \int_{\mathbb{R}^d} Mz \cdot \nabla\rho(z) \, dz \right|$$

$$= \left| -\int_{\mathbb{R}^d} \operatorname{div}(Mz)\rho(z) \, dz \right|$$

$$= \left| -\operatorname{trace}(M) \int_{\mathbb{R}^d} \rho(z) \, dz \right| = |\operatorname{trace}(M)|.$$

We pass to the opposite inequality. Since

$$Mz \cdot \nabla\rho(z) = \operatorname{div}(Mz\,\rho(z)) - \operatorname{trace}(M)\,\rho(z)$$

and $\rho \geq 0$, we immediately obtain

$$\int_{\mathbb{R}^d} |Mz \cdot \nabla\rho(z)| \, dz \leq \int_{\mathbb{R}^d} |\operatorname{div}(Mz\,\rho(z))| + |\operatorname{trace}(M)|\,\rho(z)\,dz$$

$$= \int_{\mathbb{R}^d} |\operatorname{div}(Mz\,\rho(z))| \, dz + |\operatorname{trace}(M)|.$$

Then it suffices to show that

$$\inf_{\rho \in \mathcal{K}} \int_{\mathbb{R}^d} |\operatorname{div}(Mz\,\rho(z))| \, dz = 0. \tag{2.32}$$

STEP 1. We preliminarily show that

$$\inf_{\mu} \int_{\mathbb{R}^d} |\mathrm{div}\,(Mz\,\mu)| = 0, \qquad (2.33)$$

where the infimum is now taken over all probability measures μ on $\mathbb{R}^d$. With the integral we intend the total variation of the distribution $\mathrm{div}\,(Mz\,\mu)$, i.e.

$$\sup\left\{ \langle Mz\,\mu, \nabla\varphi \rangle \;:\; \varphi \in C_c^\infty(\mathbb{R}^d), \quad \|\varphi\|_\infty \le 1 \right\}.$$

In particular, the total variation is finite if and only if $\mathrm{div}\,(Mz\mu)$ is a finite measure, and if this is the case it coincides with $\|\mathrm{div}\,(Mz\,\mu)\|_{\mathcal{M}(\mathbb{R}^d)}$.

Without loss of generality, we can assume $M \ne 0$. Now let us consider $\mu = \xi \mathcal{H}^1 \llcorner \Gamma$, where Γ is the support of a nonconstant integral line $\gamma : [0, T] \to \mathbb{R}^d$ of the vector field Mz, i.e. γ satisfies $\dot{\gamma}(t) = M\gamma(t)$ for every $t \in [0, T]$. The density ξ is of the form

$$\xi(z) = \#\{\gamma^{-1}(z)\}\frac{\alpha}{|Mz|},$$

where α is the normalization constant

$$\alpha^{-1} = \int_\Gamma \frac{\#\{\gamma^{-1}(z)\}}{|Mz|}\,d\mathcal{H}^1(z) = \int_0^T \frac{1}{|M\gamma(t)|}|\dot{\gamma}(t)|\,dt = T.$$

Since

$$\begin{aligned}
\langle \mathrm{div}\,(Mz\,\mu), \varphi \rangle &= -\langle Mz\,\mu, \nabla\varphi \rangle = -\int_{\mathbb{R}^d} Mz \cdot \nabla\varphi(z)\,d\mu(z) \\
&= -\int_\Gamma Mz \cdot \nabla\varphi(z)\frac{\#\{\gamma^{-1}(z)\}}{T|Mz|}\,d\mathcal{H}^1(z) \\
&= -\frac{1}{T}\int_0^T \dot{\gamma}(t) \cdot \nabla\varphi(\gamma(t))\,dt \\
&= -\frac{1}{T}[\varphi(\gamma(T)) - \varphi(\gamma(0))]
\end{aligned}$$

for every $\varphi \in C_c^\infty(\mathbb{R}^d)$, we get

$$\mathrm{div}\,(Mz\,\mu) = -\frac{1}{T}[\delta_{\gamma(T)} - \delta_{\gamma(0)}] \qquad \text{in } \mathcal{D}'(\mathbb{R}^d).$$

Hence

$$\int_{\mathbb{R}^d} |\mathrm{div}\,(Mz\,\mu)| \le \frac{2}{T}$$

for any such μ. Then (2.33) follows immediately. Indeed, taken any maximal integral line $\gamma : [0, +\infty[\rightarrow \mathbb{R}^d$, it is enough to consider its restriction to $[0, T]$ and then to use the arbitrariness of the choice of T.

STEP 2. Now we show the validity of (2.32). For every $x \in \mathbb{R}^d$ we consider the maximal solution $\gamma_x : [0, +\infty[\rightarrow \mathbb{R}^d$ of the problem

$$\begin{cases} \dot{\gamma}_x(t) = M\gamma_x(t) \\ \gamma_x(0) = x . \end{cases}$$

It exists for every initial point $x \in \mathbb{R}^d$ and is explicitly given by the expression $\gamma_x(t) = e^{tM}x$. We define

$$\mathcal{A} = \left\{ x \in \mathbb{R}^d : \gamma_x \text{ is not a constant} \right\} .$$

Notice that $\mathcal{A}$ is a nonempty open set which is symmetrical with respect to the origin. Now fix an even and positive function $\theta \in C_c^\infty(\mathbb{R}^d)$ with $\int_{\mathbb{R}^d} \theta = 1$ and such that $\mathrm{spt}\, \theta \subset \mathcal{A}$. We set

$$\nu^T = \int_{\mathbb{R}^d} \mu_x^T \theta(x)\, dx ,$$

with μ_x^T defined as before, starting from $\gamma_x|_{[0,T]}$.

Let us check that ν^T is in fact a smooth function. If $\varphi \in C_c(\mathbb{R}^d)$ we have

$$\langle \mu_x^T, \varphi \rangle = \frac{1}{T} \int_0^T \varphi(\gamma_x(t))\, dt .$$

Hence we deduce the following expression for ν^T:

$$\langle \nu^T, \varphi \rangle = \int_{\mathbb{R}^d} \langle \mu_x^T, \varphi \rangle \theta(x)\, dx = \frac{1}{T} \int_0^T \int_{\mathbb{R}^d} \varphi(e^{tM}x)\theta(x)\, dx dt$$

$$= \frac{1}{T} \int_0^T \int_{\mathbb{R}^d} \varphi(y)\theta(e^{-tM}y) \det(e^{-tM})\, dy dt$$

$$= \int_{\mathbb{R}^d} \left(\frac{1}{T} \int_0^T \theta(e^{-tM}y) \det(e^{-tM})\, dt \right) \varphi(y)\, dy .$$

It is then clear that ν^T is a smooth function which satisfies all the constraints of the problem (2.32). Moreover

$$\mathrm{div}_z\, (Mz\, \nu^T(z)) = \int_{\mathbb{R}^d} \mathrm{div}_z\, (Mz\, \mu_x^T)\theta(x)\, dx ,$$

where the divergence inside the integral is in the sense of distributions, for every fixed value of x. Eventually we get

$$\int_{\mathbb{R}^d} |\text{div}_z \left(Mz\, v^T (z) \right)|\, dz \leq \frac{2}{T}.$$

From the arbitrariness of T we obtain (2.32).   □

2.7. Other various renormalization results

We now briefly indicate some other renormalization results for vector fields with some particular structure. We will not enter into the proofs of such results but we will rather give some basic references on each of them.

2.7.1. Conditions on the symmetric part of the derivative

In [50] Capuzzo Dolcetta and Perthame show, among other things, that the renormalization property holds for vector fields b such that the *symmetric derivative*

$$Eb = \frac{1}{2}\left(Db + {}^t Db \right) \in \left[\mathcal{D}'(I \times \mathbb{R}^d) \right]^{d \times d}$$

is absolutely continuous with respect to the Lebesgue measure. Their proof is an adaptation of DiPerna–Lions' one: if the convolution kernel ρ used in the regularization procedure is chosen to be radial then it is simple to see that, for every direction $z \in \mathbb{R}^d$, in order to control the commutator it is not necessary to control the whole difference quotient in the direction z, but only its scalar product with z itself. This control is precisely the one ensured by the condition on the symmetric derivative.

It would be interesting to adapt this proof to show the renormalization property for vector fields with *bounded deformation, i.e.* such that the symmetric part of the derivative is a measure (see Appendix A.3 for a more precise definition of the space BD). However, as we have just observed, conditions on the symmetric part of the derivative seem exploitable only through the use of radial convolution kernels; hence the whole anisotropic regularization technique is apparently not disposable in this context.

A result in this direction has been obtained in a work in collaboration with Ambrosio and Maniglia [14]: we are able to treat the case of *special vector fields with bounded deformation*, which satisfy by assumption the condition that the singular part of the symmetric derivative is concentrated on a set of codimension one. The singular part is ruled out using a chain-rule formula for normal traces of vector fields with measure divergence. See also [25, 55, 56] for general results on this notion of trace. However the Cantor part of the derivative is not treatable with this kind of techniques.

2.7.2. Partial regularity of the vector field

Looking at the results by DiPerna and Lions and by Ambrosio it is apparent that the regularity needed on b in the dependence on the spatial variable (Sobolev or BV) is much greater than the one needed in the dependence on the time variable (just summability is requested). If we look at the vector field $B(t, x) = (1, b(t, x))$ in space-time, for which the transport equation reads

$$B(t, x) \cdot \nabla_{t,x} u(t, x) = 0,$$

we see that B has a kind of "split structure", in which the first component depends fairly on the first variable, and the second component depends fairly on the second variable, but very wildly on the first one. It is then natural to conjecture (see [8, Remark 3.8]) that the renormalization property should hold for vector fields which admit a splitting of the type

$$B = \big(b_1(x_1), b_2(x_1, x_2)\big), \qquad x_1, b_1 \in \mathbb{R}^{d_1}, \qquad x_2, b_2 \in \mathbb{R}^{d_2},$$

under an assumption of good dependence of b_1 on x_1 and of b_2 on x_2. Some results in this direction are available in [104, 105, 106], in which the partial regularity is of Sobolev or BV type. The proof relies again on an anisotropic convolution: the idea is that we have to regularize faster along the x_1 direction, because b_2 has no regularity in that direction. It would be interesting to understand if results of this kind can hold if we admit also an x_2 dependence of the b_1 component.

2.7.3. The two-dimensional case

Consider a bounded autonomous vector field $b : \mathbb{R}^2 \to \mathbb{R}^2$ with $\operatorname{div} b = 0$. Then it is possible to find a Lipschitz function $H : \mathbb{R}^2 \to \mathbb{R}$ (called the Hamiltonian) such that $b = \nabla^{\perp} H$. Heuristically we expect that the evolution is split on the level lines of H, since the value of H is conserved under the flow. Hence we expect a kind of dimensional reduction of the problem, which we could try to solve "line by line". However, in this very general setting, the well-posedness does not hold: in a joint work with Alberti and Bianchini [5] we construct counterexamples to the uniqueness for bounded autonomous planar divergence-free vector fields. For the uniqueness to hold we need to require a weak Sard property of the

Hamiltonian (see (4.11) and Theorem 4.5.2), under which we are able to implement the idea of the splitting on the level curves of H. This result has been proved in the same paper [5] and will be presented in Chapter 4, where we will also indicate more references on the problem.

2.8. Nearly incompressible vector fields

The aim of this section is to develop a well-posedness theory for vector fields in which the usual assumption of boundedness of the divergence is replaced by a control of the Jacobian, or by the existence of a solution of the continuity equation which is bounded away from zero and infinity. This is particularly important in view of the applications, for instance to the Keyfitz and Kranzer system (see [15, 11]). A systematic treatement of this topic is done in [77]. Since the assumption on the divergence is necessary in order to give a distributional meaning to the transport equation (see Section 2.2), we deal here with the equation in continuity form. We will consider again the case of nearly incompressible vector fields, from the ODE viewpoint, in Sections 6.6 and 6.7.

Definition 2.8.1 (Nearly incompressible vector fields). We say that a bounded vector field $b : I \times \mathbb{R}^d \to \mathbb{R}^d$ is *nearly incompressible* if there exist a function $\rho \in L^\infty(I \times \mathbb{R}^d)$ and a constant $C > 0$ such that

$$0 < \frac{1}{C} \le \rho(t, x) \le C < +\infty \qquad \text{for } \mathscr{L}^{d+1}\text{-a.e. } (t, x) \in I \times \mathbb{R}^d$$
(2.34)

and

$$\partial_t \rho + \operatorname{div}(b\rho) = 0 \qquad \text{in } \mathcal{D}'(I \times \mathbb{R}^d).$$
(2.35)

Notice that, thanks to Remark 2.2.2, we can always assume $\rho \in C([0, T]; L^\infty(\mathbb{R}^d) - w^*)$. We remark that every vector field b with bounded divergence is nearly incompressible: if b is smooth, take for ρ the Jacobian determinant of the flow generated by b, that is $\rho(t, x) = \det \nabla_x X(0, t, x)$, which is bounded since we can compute explicitly

$$\rho(t, x) = \exp\Big[-\int_0^t (\operatorname{div} b)(\sigma, X(\sigma, t, x))d\sigma\Big],$$

where $X(s,t,x)$ satisfies $\partial X(s,t,x)/\partial s = b(s, X(s,t,x))$ and $X(t,t,x) = x$. A standard approximation argument settles the case of non-smooth vector fields. In general, however, a nearly incompressible vector field does not need to have absolutely continuous divergence.

In the spirit of the discussion of Section 2.3 we introduce the definition of renormalization property in this new context.

Definition 2.8.2. (Renormalization property for nearly incompress-ible vector fields). We say that a bounded nearly incompressible vector field $b : I \times \mathbb{R}^d \to \mathbb{R}^d$ has the *renormalization property* if, for some function ρ as in Definition 2.8.1, every solution $u \in L^\infty(I \times \mathbb{R}^d)$ of

$$\partial_t(\rho u) + \mathrm{div}\,(b\rho u) = 0 \qquad \text{in } \mathcal{D}'(I \times \mathbb{R}^d)$$

satisfies

$$\partial_t(\rho\beta(u)) + \mathrm{div}\,(b\rho\beta(u)) = 0 \qquad \text{in } \mathcal{D}'(I \times \mathbb{R}^d)$$

for every function $\beta \in C^1(\mathbb{R}; \mathbb{R})$.

It can be checked that the renormalization property is independent of the choice of the function ρ in Definition 2.8.1. For bounded nearly incompressible vector fields with the renormalization property it is possible to develop a well-posedness theory arguing as in Section 2.3: the following proposition can be proved with the same techniques used there.

Proposition 2.8.3. *Let* $b : [0, T] \times \mathbb{R}^d \to \mathbb{R}^d$ *be a bounded nearly incompressible vector field and let* ρ *as in Definition* 2.8.1. *Define the vector field* $\tilde{b} :]-\infty, T] \times \mathbb{R}^d \to \mathbb{R}^d$ *according to*

$$\tilde{b}(t, x) = \begin{cases} 0 & \text{if } t < 0 \\ b(t, x) & \text{if } 0 \le t \le T \end{cases} \qquad (2.36)$$

and assume that $\tilde{b}$ *has the renormalization property in the sense of Definition* 2.8.2. *Then for every* $\bar{u} \in L^\infty(\mathbb{R}^d)$ *there exists a unique bounded solution* u *of*

$$\begin{cases} \partial_t(\rho u) + \mathrm{div}\,(b\rho u) = 0 \\ (\rho u)(0, \cdot) = \rho(0, \cdot)\bar{u}\,. \end{cases}$$

This result implies uniqueness for the continuity equation for nearly incompressible vector fields with the renormalization property. Some stability results are also available, for which we refer to [77, Section 3.4].

Theorem 2.8.4. *Let* $b : [0, T] \times \mathbb{R}^d \to \mathbb{R}^d$ *be a bounded nearly incompressible vector field and assume that the vector field* $\tilde{b}$ *defined as in* (2.36) *has the renormalization property in the sense of Definition* 2.8.2. *Then for every* $\bar{\zeta} \in L^\infty(\mathbb{R}^d)$ *there exists a unique solution* $\zeta \in L^\infty([0,T] \times \mathbb{R}^d)$ *to*

$$\begin{cases} \partial_t\zeta + \mathrm{div}\,(b\zeta) = 0 \\ \zeta(0, \cdot) = \bar{\zeta}\,. \end{cases}$$

The previous uniqueness result justifies the following definition, which will be important in the discussion of Section 6.6.

Definition 2.8.5 (Density generated by a vector field). Let $b : [0, T] \times \mathbb{R}^d \to \mathbb{R}^d$ be a bounded nearly incompressible vector field such that the vector field $\tilde{b}$ defined as in (2.36) has the renormalization property in the sense of Definition 2.8.2. Then the *density generated by* b is the unique solution $\rho \in L^\infty([0, T] \times \mathbb{R}^d)$ of

$$\begin{cases} \partial_t \rho + \operatorname{div}(b\rho) = 0 \\ \rho(0, \cdot) = 1 . \end{cases}$$

Chapter 3
An abstract characterization of the renormalization property

In this chapter we present a joint work with Bouchut [37], in which we give a result of different type compared with the ones of the previous chapter. We do not want to give new well-posedness theorems, but rather equivalent conditions for the well-posedness to hold, without regularity assumptions on b.

For simplicity we shall always assume that $b \in L^\infty([0, T] \times \mathbb{R}^d)$, and consider an L^2 framework. The approach of [84] and [8] illustrated in Sections 2.5 and 2.6 relies on an approximation by convolution of a given weak solution to the transport equation (2.1) and on the renormalization property. Our first theorem (see Theorem 3.1.1) states that such properties are indeed equivalent to the well-posedness of both forward and backward Cauchy problems, up to the fact that the smooth approximate solution (in the sense of the norm of the graph of the transport operator) is not necessarily given by convolution. See also [33] for some related results. This is also consistent with the preliminary discussion of Section 2.1, in which we observed the importance of the renormalization property and of the strong continuity of the solution with respect to the time: see again Theorem 3.1.1 and also Remark 3.1.7.

Then, one can consider separately the two different issues of forward and backward uniqueness. Theorem 3.2.1 states that a characterization of backward uniqueness is the existence of a solution to the forward Cauchy problem that is approximable by smooth functions in the sense of the norm of the graph of the transport operator. Finally, we are also able to consider the case of nearly incompressible vector fields, with possibly unbounded divergence. We show that the previous results extend naturally to this case.

3.1. Forward-backward formulation

We will be concerned with the continuity equation

$$\partial_t u + \operatorname{div}(bu) = 0 \quad \text{in } \mathcal{D}'([0, T] \times \mathbb{R}^d), \tag{3.1}$$

where the vector field $b : [0, T] \times \mathbb{R}^d \to \mathbb{R}^d$ is bounded. No regularity is assumed on b. We start, in the divergence-free context, with a characterization result which relates renormalization, strong continuity, uniqueness and approximation with smooth functions of the solution. Notice that, in the divergence-free case, the continuity equation and the transport equation coincide.

Theorem 3.1.1. *Let $b \in L^\infty([0, T] \times \mathbb{R}^d; \mathbb{R}^d)$ such that $\operatorname{div} b = 0$. Then the following statements are equivalent:*

(i) *b has the uniqueness property for weak solutions in $C([0,T]; L^2(\mathbb{R}^d)-w)$ for both the forward and the backward Cauchy problems starting, respectively, from 0 and T; i.e., the only solutions in $C([0,T]; L^2(\mathbb{R}^d)-w)$ to the problems*

$$\begin{cases} \partial_t u_F + \operatorname{div}(bu_F) = 0 \\ u_F(0, \cdot) = 0, \end{cases} \qquad \begin{cases} \partial_t u_B + \operatorname{div}(bu_B) = 0 \\ u_B(T, \cdot) = 0 \end{cases}$$

are $u_F \equiv 0$ and $u_B \equiv 0$.

(ii) *The Banach space*

$$\mathcal{F} = \left\{ \begin{array}{l} u \in C([0, T]; L^2(\mathbb{R}^d) - w) \text{ such that} \\ \partial_t u + \operatorname{div}(bu) \in L^2([0, T] \times \mathbb{R}^d) \end{array} \right\} \tag{3.2}$$

with norm

$$\|u\|_{\mathcal{F}} = \|u\|_{B([0,T]; L^2(\mathbb{R}^d))} + \|\partial_t u + \operatorname{div}(bu)\|_{L^2([0,T] \times \mathbb{R}^d)} \tag{3.3}$$

has the property that the space of functions in $C^\infty([0, T] \times \mathbb{R}^d)$ with compact support in x is dense in $\mathcal{F}$.

(iii) *Every weak solution in $C([0, T]; L^2(\mathbb{R}^d) - w)$ of $\partial_t u + \operatorname{div}(bu) = 0$ lies in $C([0, T]; L^2(\mathbb{R}^d) - s)$ and is a renormalized solution, i.e., for every function $\beta \in C^1(\mathbb{R}; \mathbb{R})$ such that $|\beta'(s)| \leq C(1 + |s|)$ for some constant $C \geq 0$, one has $\partial_t(\beta(u)) + \operatorname{div}(b\beta(u)) = 0$ in $[0, T] \times \mathbb{R}^d$.*

In the statement of the theorem we used the notation $C([0, T]; L^2(\mathbb{R}^d) - w)$ and $C([0, T]; L^2(\mathbb{R}^d) - s)$ for the spaces of maps which are continuous from $[0, T]$ into $L^2(\mathbb{R}^d)$, endowed with the weak and the strong topology, respectively. With $B([0, T]; L^2(\mathbb{R}^d))$ we denoted the space of bounded maps from $[0, T]$ into $L^2(\mathbb{R}^d)$, intending that maps in this space are defined for *every* $t \in [0, T]$. We recall that, up to a redefinition in a negligible set of times, every solution to (3.1) belongs to $C([0, T]; L^2(\mathbb{R}^d) - w)$ (see Remark 2.2.2).

Proof of Theorem 3.1.1. (I) $\Rightarrow$ (II)

STEP 1. CAUCHY PROBLEM IN $\mathcal{F}$. It is easy to check that $\mathcal{F}$ is a Banach space, since $L^2([0, T] \times \mathbb{R}^d)$ and $B([0, T]; L^2(\mathbb{R}^d))$ are Banach spaces (the latter denotes the space of bounded functions, with the supremum norm). We preliminarily show that for any $f \in L^2([0, T] \times \mathbb{R}^d)$ and $u^0 \in L^2(\mathbb{R}^d)$, the Cauchy problem

$$\begin{cases} \partial_t u + \operatorname{div}(bu) = f \\ u(0, \cdot) = u^0 \end{cases} \tag{3.4}$$

has a unique solution in $\mathcal{F}$. We proceed by regularization. Consider a sequence of smooth vector fields $\{b_n\}$, with $b_n \to b$ a.e., b_n uniformly bounded in L^∞, and $\operatorname{div} b_n = 0$ for every n. Let u_n be the solution to the problem

$$\begin{cases} \partial_t u_n + \operatorname{div}(b_n u_n) = f \\ u_n(0, \cdot) = u^0. \end{cases}$$

We know that the solution u_n is unique in $C([0, T]; L^2(\mathbb{R}^d))$ and is given by

$$u_n(t, x) = u^0(X_n(0, t, x)) + \int_0^t f(\tau, X_n(\tau, t, x)) \, d\tau$$

(see formula (1.16)), where $X_n(s, t, x)$ is as usual the flow of b_n at time s, starting at the point x at time t. Recalling that $\operatorname{div} b_n = 0$, so that $X_n(s, t, \cdot)_\# \mathscr{L}^d = \mathscr{L}^d$ for every s and t, we can estimate the L^2 norm of $u_n(t, \cdot)$ as follows:

$$\|u_n(t, \cdot)\|_{L^2} \leq \|u^0(X_n(0, t, \cdot))\|_{L^2} + \int_0^t \|f(\tau, X_n(\tau, t, \cdot))\|_{L^2} \, d\tau$$

$$\leq \|u^0\|_{L^2} + \int_0^t \|f(\tau, \cdot)\|_{L^2} \, d\tau$$

$$\leq \|u^0\|_{L^2} + \sqrt{T} \|f\|_{L^2}.$$

This implies that the sequence $\{u_n\}$ is equi-bounded in $C([0, T]; L^2(\mathbb{R}^d))$. From the equation on u_n, we have also that, for any $\varphi \in C_c^\infty(\mathbb{R}^d)$, $d/dt(\int u_n \varphi \, dx)$ is bounded in $L^2[0, T]$. We deduce that, for any $\varphi \in L^2(\mathbb{R}^d)$, $\int u_n \varphi \, dx$ is equicontinuous in $[0, T]$ uniformly in n. Thus, up to subsequences, we can suppose that $u_n \to u$ in $C([0, T]; L^2(\mathbb{R}^d) - w)$. By the semicontinuity of the norm with respect to weak convergence we also obtain that

$$\|u(t, \cdot)\|_{L^2} \leq \|u^0\|_{L^2} + \sqrt{T} \|f\|_{L^2}. \tag{3.5}$$

Passing to the limit in the continuity equation, we obtain that u solves the Cauchy problem

$$\begin{cases} \partial_t u + \operatorname{div}(bu) = f \\ u(0, \cdot) = u^0. \end{cases}$$

Noticing that $\partial_t u + \operatorname{div}(bu) = f \in L^2$, we conclude that $u \in \mathcal{F}$. Uniqueness is clear: every solution to the Cauchy problem (3.4) is by definition a weak solution in $C([0, T]; L^2(\mathbb{R}^d) - w)$ of the forward Cauchy problem with right-hand side, and thus, by linearity, uniqueness is guaranteed by the forward part of assumption (i).

STEP 2. DENSITY OF SMOOTH FUNCTIONS. Define a linear operator

$$A: \begin{aligned} \mathcal{F} &\to L^2(\mathbb{R}^d) \times L^2([0, T] \times \mathbb{R}^d), \\ u &\mapsto \big(u(0, \cdot), \ \partial_t u + \operatorname{div}(bu)\big). \end{aligned}$$

This operator is clearly bounded by the definition of the norm we have taken on $\mathcal{F}$. It is also a bijection because of Step 1, with continuous inverse because of (3.5). This means that A is an isomorphism, and thus we can identify $\mathcal{F}$ with the space $L^2(\mathbb{R}^d) \times L^2([0, T] \times \mathbb{R}^d)$, and its dual $\mathcal{F}^*$ with $L^2(\mathbb{R}^d) \times L^2([0, T] \times \mathbb{R}^d)$. Therefore, for every functional $L \in \mathcal{F}^*$, we can uniquely define $v_0 \in L^2(\mathbb{R}^d)$ and $v \in L^2([0, T] \times \mathbb{R}^d)$ in such a way that

$$Lu = \int_{[0,T] \times \mathbb{R}^d} \big(\partial_t u + \operatorname{div}(bu)\big) v \, dt dx + \int_{\mathbb{R}^d} u(0, \cdot) v_0 \, dx \quad \text{for every } u \in \mathcal{F}.$$

We recall the classical fact that a subspace of a Banach space is dense if and only if every functional which is zero on the subspace is in fact identically zero. Then the density of smooth functions is equivalent to the following implication:

$$\left\{ \begin{aligned} &\int_{[0,T] \times \mathbb{R}^d} \big(\partial_t u + \operatorname{div}(bu)\big) v \, dt dx + \int_{\mathbb{R}^d} u(0, \cdot) v_0 \, dx = 0 \\ &\forall u \in C^\infty([0, T] \times \mathbb{R}^d) \text{ with compact support in } x \end{aligned} \right\} \tag{3.6}$$

$$\implies \left\{ \begin{aligned} v_0 &= 0 \\ v &= 0 \end{aligned} \right\}.$$

If we first take u arbitrary but with compact support also in time, we obtain that

$$\int_{[0,T] \times \mathbb{R}^d} \big(\partial_t u + \operatorname{div}(bu)\big) v \, dt dx = 0,$$

and since div $b = 0$, this is precisely the weak form of

$$\partial_t v + \text{div} (bv) = 0.$$

This implies that $v \in C([0, T]; L^2(\mathbb{R}^d) - w)$. Now, let χ be a cut-off function on $\mathbb{R}$, i.e., $\chi \in C_c^\infty(\mathbb{R})$, $\chi(z) = 1$ for $|z| \leq 1$, and $\chi(z) = 0$ for $|z| \geq 2$. For every function $\varphi \in C_c^\infty(\mathbb{R}^d)$, take a function $\tilde{u} \in C^\infty([0, T] \times \mathbb{R}^d)$ with compact support in x such that $\tilde{u}(T, \cdot) = \varphi$. Then, testing in (3.6) with $u(t, x) = \tilde{u}(t, x)\chi\big((T - t)/\epsilon\big)$, we obtain for $0 < \epsilon < T/2$

$$0 = \int_{[0,T]\times\mathbb{R}^d} \left[\partial_t \left(\tilde{u}(t, x)\chi\left(\frac{T-t}{\epsilon}\right)\right) \right.$$

$$\left. + \text{div}\left(b(t, x)\tilde{u}(t, x)\chi\left(\frac{T-t}{\epsilon}\right)\right)\right] v(t, x)\,dt\,dx$$

$$\tag{3.7}$$

$$= \int_{[0,T]\times\mathbb{R}^d} \left[\partial_t\tilde{u}(t, x) + \text{div}\,(b(t,x)\tilde{u}(t,x))\right] v(t,x)\chi\left(\frac{T-t}{\epsilon}\right)dt\,dx$$

$$- \int_{[0,T]\times\mathbb{R}^d} \frac{1}{\epsilon}\chi'\left(\frac{T-t}{\epsilon}\right) \tilde{u}(t, x)v(t, x)\,dt\,dx.$$

Letting $\epsilon \to 0$, we observe that the first integral converges to 0, since the support of $\chi\big((T - t)/\epsilon\big)$ is contained in $[T - 2\epsilon, T + 2\epsilon]$. The second integral can be rewritten as

$$- \int_0^T \frac{1}{\epsilon}\chi'\left(\frac{T-t}{\epsilon}\right)\left[\int_{\mathbb{R}^d} \tilde{u}(t, x)v(t, x)\,dx\right]dt.$$

Now, since $\tilde{u}$ is smooth and $v \in C([0, T]; L^2(\mathbb{R}^d) - w)$, the integral over $\mathbb{R}^d$ is a continuous function of t. Moreover, it is easy to check that

$$- \int_0^T \frac{1}{\epsilon}\chi'\left(\frac{T-t}{\epsilon}\right)dt = 1.$$

Therefore, coming back to (3.7) and letting $\epsilon \to 0$ we get

$$0 = \int_{\mathbb{R}^d} \tilde{u}(T, x)v(T, x)\,dx = \int_{\mathbb{R}^d} \varphi(x)v(T, x)\,dx.$$

Since $\varphi \in C_c^\infty(\mathbb{R}^d)$ is arbitrary, we obtain $v(T, \cdot) = 0$. We conclude that $v \in C([0, T]; L^2(\mathbb{R}^d) - w)$ solves the Cauchy problem

$$\begin{cases} \partial_t v + \text{div}\,(bv) = 0 \\ v(T, \cdot) = 0. \end{cases}$$

Thus, by the backward part of the uniqueness assumption (i), we get that $v = 0$. Substituting in (3.6), we get that $\int_{\mathbb{R}^d} u(0, \cdot)v_0\, dx = 0$ for every $u \in C^\infty([0, T] \times \mathbb{R}^d)$ with compact support in space, and this implies that $v_0 = 0$. This concludes the proof of the implication (3.6), which ensures that (ii) holds.

(II) $\Rightarrow$ (III)

Let $u \in C([0, T], L^2(\mathbb{R}^d) - w)$ satisfy $\partial_t u + \operatorname{div}(bu) = 0$. Then, by (ii) there exists a sequence $\{u_n\}$ of functions in $C^\infty([0, T] \times \mathbb{R}^d)$ with compact support in space such that $\|u_n - u\|_{\mathcal{F}} \to 0$. In particular this gives that $u_n \to u$ in $B([0, T]; L^2(\mathbb{R}^d))$; thus $u \in C([0, T], L^2(\mathbb{R}^d) - s)$. Then, define $f_n = \partial_t u_n + \operatorname{div}(bu_n) \in L^2([0, T] \times \mathbb{R}^d)$. By the definition of convergence in $\mathcal{F}$ we have that $f_n \to 0$ strongly in $L^2([0, T] \times \mathbb{R}^d)$. For every function β with the regularity stated we can apply the classical chain-rule, giving

$$\partial_t(\beta(u_n)) + \operatorname{div}(b\beta(u_n)) = \beta'(u_n)f_n. \tag{3.8}$$

The left-hand side clearly converges to $\partial_t(\beta(u)) + \operatorname{div}(b\beta(u))$ in the sense of distributions. Since $|\beta'(s)| \leq C(1 + |s|)$, the sequence $\beta'(u_n)$ is equi-bounded in $L^2_{\text{loc}}([0, T] \times \mathbb{R}^d)$; hence from the strong convergence of f_n we deduce that the right-hand side of (3.8) converges strongly to zero in $L^1_{\text{loc}}([0, T] \times \mathbb{R}^d)$. This implies that $\partial_t(\beta(u)) + \operatorname{div}(b\beta(u)) = 0$.

(III) $\Rightarrow$ (I)

This implication is very similar to the one given in Theorem 2.3.3. However, we present it with some detail, mostly in order to enlighten the role of the strong continuity assumption.

Let $u \in C([0, T], L^2(\mathbb{R}^d) - w)$ satisfy $\partial_t u + \operatorname{div}(bu) = 0$. According to (iii), u lies in $C([0, T], L^2(\mathbb{R}^d) - s)$, and applying the renormalization property with $\beta(u) = u^2$, we get

$$\partial_t u^2 + \operatorname{div}(bu^2) = 0, \tag{3.9}$$

with $u^2 \in C([0, T], L^1(\mathbb{R}^d) - s)$. Consider $\psi \in C_c^\infty([0, T])$ and $\varphi_R(x) = \varphi(x/R)$, where $\varphi \in C_c^\infty(\mathbb{R}^d)$ is a cut-off function equal to 1 on the ball of radius 1 and equal to 0 outside the ball of radius 2. Testing equation (3.9) against functions of the form $\psi(t)\varphi_R(x)$ we get

$$\int_0^T \left[\int_{\mathbb{R}^d} u^2 \varphi\left(\frac{x}{R}\right) dx \right] \psi'(t)dt + \int_0^T \left[\int_{\mathbb{R}^d} bu^2 \frac{1}{R}\nabla\varphi\left(\frac{x}{R}\right) dx \right] \psi(t)dt = 0.$$

Thus,

$$\frac{d}{dt} \int_{\mathbb{R}^d} u^2 \varphi\left(\frac{x}{R}\right) dx = \int_{\mathbb{R}^d} bu^2 \frac{1}{R}\nabla\varphi\left(\frac{x}{R}\right) dx \quad \text{in } \mathcal{D}'([0, T]). \tag{3.10}$$

Since the right-hand side of (3.10) is in $L^\infty([0, T])$ and since for every t it is bounded by the quantity $\frac{1}{R}\|b\|_{L^\infty_{t,x}}\|\nabla\varphi\|_{L^\infty_x}\|u(t,\cdot)\|^2_{L^2_x}$, letting $R \to +\infty$ we obtain

$$\frac{d}{dt}\int_{\mathbb{R}^d} u(t,x)^2\,dx = 0 \quad \text{in } [0, T]. \tag{3.11}$$

Recalling that $u^2 \in C([0, T], L^1(\mathbb{R}^d) - s)$, (3.11) yields $\int u(t,x)^2 dx = cst$ on $[0, T]$, which implies uniqueness for both the forward and the backward Cauchy problems, proving (i). $\qquad\square$

Remark 3.1.2 (Well-posedness). The space $\mathcal{F}$ defined in (3.2) is a natural space for the study of the Cauchy problem (3.4). Whenever one of the statements of Theorem 3.1.1 is true, we have existence and uniqueness in $\mathcal{F}$ with the estimate (3.5), as shown in the proof. Moreover, every solution is renormalized and strongly continuous with respect to time, *i.e.*, $u \in C([0, T]; L^2(\mathbb{R}^d) - s)$. Overall, the following weak stability holds: if

- $\{f_n\}$ is a bounded sequence in $L^2([0,T]\times\mathbb{R}^d)$ which converges weakly to f,
- $\{u^0_n\}$ is a bounded sequence in $L^2(\mathbb{R}^d)$ which converges weakly to u^0,
- $\{b_n\}$ is a bounded sequence in $L^\infty([0, T] \times \mathbb{R}^d)$ which converges strongly in L^1_{loc} to b and such that $\operatorname{div} b_n = 0$ for every n,

then the solutions $\{u_n\}$ to

$$\partial_t u_n + \operatorname{div}(b_n u_n) = f_n, \qquad u_n(0,\cdot) = u^0_n$$

converge in $C([0, T]; L^2(\mathbb{R}^d) - w)$ to the solution u to the Cauchy problem (3.4).

Remark 3.1.3 (L^p case). We can modify the summability exponent in the definition of the space $\mathcal{F}$. For every $p \in]1, \infty[$, define $\mathcal{F}_p$ as the space containing those functions $u \in C([0, T]; L^p(\mathbb{R}^d) - w)$ that satisfy $\partial_t u + \operatorname{div}(bu) \in L^p([0, T]\times\mathbb{R}^d)$ and define the norm $\|\cdot\|_{\mathcal{F}_p}$ in the obvious way, which makes $\mathcal{F}_p$ a Banach space. Denoting by p' the conjugate exponent of p, *i.e.*, $\frac{1}{p} + \frac{1}{p'} = 1$, the following statements are equivalent:

(i) Smooth functions with compact support in x are dense in $\mathcal{F}_p$ and in $\mathcal{F}_{p'}$.
(ii) The vector field b has the forward uniqueness property for weak solutions in $C([0, T]; L^p(\mathbb{R}^d) - w)$ and the backward uniqueness property for weak solutions in $C([0, T]; L^{p'}(\mathbb{R}^d) - w)$.

Remark 3.1.4 (Equivalent norms). According to the proof of Theorem 3.1.1, if one of the properties (i), (ii), and (iii) holds, then the norm of $\mathcal{F}$ is equivalent to the norm

$$\|u\|_{\mathcal{F},0} = \|u(0,\cdot)\|_{L^2(\mathbb{R}^d)} + \|\partial_t u + \mathrm{div}\,(bu)\|_{L^2([0,T]\times\mathbb{R}^d)}$$

(see the estimate (3.5)). In the same spirit, it is easy to prove that $\|\cdot\|_{\mathcal{F}}$ is in fact equivalent to every norm of the form

$$\|u\|_{\mathcal{F},s} = \|u(s,\cdot)\|_{L^2(\mathbb{R}^d)} + \|\partial_t u + \mathrm{div}\,(bu)\|_{L^2([0,T]\times\mathbb{R}^d)}$$

for $s \in [0, T]$.

Remark 3.1.5 (Depauw's counterexample). A simple modification (translation in time) of the counterexample constructed by Depauw [82] (see Section 5.1) shows that the renormalization property is really linked to the uniqueness in *both* the forward and the backward Cauchy problems. In fact, we can construct a divergence-free vector field $b \in L^\infty([0,1] \times \mathbb{R}^2; \mathbb{R}^2)$ and a function $\bar{u} \in L^\infty(\mathbb{R}^2)$ such that

1. the backward Cauchy problem with datum $\bar{u}$ at time $t = 1$ has a unique solution, which is, however, *not* renormalized and *not* strongly continuous with respect to time;
2. the forward Cauchy problem with datum 0 at time $t = 0$ has more than one solution;
3. the unique solution $u(t, x)$ to the backward Cauchy problem with datum $\bar{u}$ at time $t = 1$ satisfies

$$\begin{cases} |u(t,x)| = 0 & \text{for } 0 \le t \le 1/2 \\ |u(t,x)| = 1 & \text{for } 1/2 < t \le 1; \end{cases}$$

hence the equivalence of the norms in Remark 3.1.4 does *not* hold.

Remark 3.1.6 (The Sobolev and the BV cases). In the case of a vector field with Sobolev regularity with respect to the space variable, $b \in L^1([0,T]; W^{1,p'}_{\mathrm{loc}}(\mathbb{R}^d))$ with $1 < p < \infty$, it is almost possible to prove that the natural regularization by convolution with respect to the space variable of $u \in \mathcal{F}_p$ (see Remark 3.1.3) converges to u with respect to $\|\cdot\|_{\mathcal{F}_p}$. Indeed, let η_ϵ be a standard convolution kernel in $\mathbb{R}^d$ and set $u_\epsilon = u * \eta_\epsilon$. We can compute

$$\partial_t u + \mathrm{div}\,(bu) - \partial_t u_\epsilon - \mathrm{div}\,(bu_\epsilon)$$
$$= \big[\partial_t u + \mathrm{div}\,(bu)\big] - \big[\partial_t u + \mathrm{div}\,(bu)\big] * \eta_\epsilon$$
$$+ \big[\mathrm{div}\,(bu) * \eta_\epsilon - \mathrm{div}\,(bu_\epsilon)\big].$$

Then the convergence of u_ϵ to u with respect to $\|\cdot\|_{\mathcal{F}_p}$ is equivalent to the strong convergence in $L^p([0, T] \times \mathbb{R}^d)$ to zero of the commutator

$$r_\epsilon = \operatorname{div}(bu) * \eta_\epsilon - \operatorname{div}(bu_\epsilon).$$

The results of [84] ensure this strong convergence for every convolution kernel η_ϵ, except that it holds in L^1_{loc} instead of L^p. We need also a regularization with respect to time and a cut-off in order to get the density property in Theorem 3.1.1(ii), but this means that our strategy is more or less "equivalent" to the one of [84], in the framework of Sobolev vector fields. However, the situation is different in the BV case studied in [8]. In general, the commutator r_ϵ is not expected to converge strongly to zero; our result shows that, even in this case, there exists some smooth approximation of the solution, but it is less clear how to construct it in an explicit way.

Remark 3.1.7 (Strong continuity condition). The condition of continuity with values in strong L^2 in Theorem 3.1.1(iii) cannot be removed; otherwise the equivalence with (i) fails. This can be seen again with Depauw's counterexample with singularity at time $t = 0$. In this case all weak solutions are renormalized in $[0, T] \times \mathbb{R}^d$ since b is locally BV in x, but uniqueness of weak solutions does not hold. Another remark is that, in general, a renormalized solution does not need to be continuous with values in strong L^2, even inside the interval, as the following counterexample shows. On the interval $[-1, 1]$, take for b Depauw's vector field in $[0, 1]$ (with singularity at 0), and define on $[-1, 0]$ the vector field as $b(t, x) = -b(-t, x)$. Consider then the weak solution u with value 0 at $t = 0$, which we extend on $[-1, 0]$ by $u(t, x) = u(-t, x)$. Then, u is a renormalized solution on $[-1, 1]$ but it is not strongly continuous at $t = 0$.

3.2. One-way formulation

We now drop the divergence-free condition, which is substituted with an L^∞ control. Moreover, we consider separately the forward and the backward problems, showing that a characterization of backward uniqueness is the existence of a solution to the forward Cauchy problem that is approximable by smooth functions in the same sense explained in Theorem 3.1.1.

Theorem 3.2.1. *Let $b \in L^\infty([0, T] \times \mathbb{R}^d; \mathbb{R}^d)$ such that $\operatorname{div} b \in L^\infty([0, T] \times \mathbb{R}^d)$, and let $c \in L^\infty([0, T] \times \mathbb{R}^d)$. Define the Banach space $\mathcal{F}$ and its norm $\|\cdot\|_{\mathcal{F}}$ as in (3.2)–(3.3). Moreover, define $\mathcal{F}^0 \subset \mathcal{F}$ as the closure (with respect to $\|\cdot\|_{\mathcal{F}}$) of the subspace of functions in $C^\infty([0, T] \times \mathbb{R}^d)$ with compact support in x. Then the following statements are equivalent:*

(i) *For every $u^0 \in L^2(\mathbb{R}^d)$ and every $f \in L^2([0, T] \times \mathbb{R}^d)$ there exists a solution $u \in \mathcal{F}^0$ to the Cauchy problem*

$$\begin{cases} \partial_t u + \mathrm{div}\,(bu) + cu = f \\ u(0, \cdot) = u^0, \end{cases} \qquad u \in \mathcal{F}^0.$$

(ii) *There is uniqueness for weak solutions in $C([0, T]; L^2(\mathbb{R}^d) - w)$ for the backward dual Cauchy problem starting from T; i.e., the only function v belonging to $C([0, T]; L^2(\mathbb{R}^d) - w)$ which solves*

$$\begin{cases} \partial_t v + b \cdot \nabla v - cv = 0 \\ v(T, \cdot) = 0 \end{cases}$$

is $v \equiv 0$.

Remark 3.2.2. The two statements in Theorem 3.2.1 are really the "non-trivial" properties relative to the vector field b. In general, there is always uniqueness in $\mathcal{F}^0$ (see Step 1 in the proof) and there is always existence of weak solutions in $\mathcal{F}$ (this can be easily proved by regularization, as in the first step of the proof of Theorem 3.1.1).

Before proving the theorem, we recall the following standard result of functional analysis (see for instance [49, Theorems II.19 and II.20]).

Lemma 3.2.3. *Let E and F be Banach spaces and let $L : E \to F$ be a bounded linear operator. Denote by $L^* : F^* \to E^*$ the adjoint operator, defined by*

$$\langle v, Lu \rangle_{F^*, F} = \langle L^* v, u \rangle_{E^*, E} \qquad \text{for every } u \in E \text{ and } v \in F^*.$$

Then

(i) *L is surjective if and only if L^* is injective and with closed image;*
(ii) *L^* is surjective if and only if L is injective and with closed image.*

Proof of Theorem 3.2.1.

STEP 1. AN ENERGY ESTIMATE IN $\mathcal{F}^0$. In this first step we prove that for every $u \in \mathcal{F}^0$ the following energy estimate holds:

$$\|u(t, \cdot)\|_{L_x^2} \leq \left(\|u(0, \cdot)\|_{L_x^2} + \sqrt{T}\|\partial_t u \right.$$

$$\left. + \mathrm{div}\,(bu) + cu\|_{L_{t,x}^2} \right) \exp \left(T \left\| c + \frac{1}{2}\mathrm{div}\,b \right\|_{L_{t,x}^\infty} \right). \tag{3.12}$$

Let us first prove the estimate for u smooth with compact support in x. We define

$$f = \partial_t u + \text{div}\,(bu) + cu,$$

and we multiply this relation by u, getting

$$\partial_t \frac{u^2}{2} + \text{div}\left(b\frac{u^2}{2}\right) + \left(c + \frac{1}{2}\text{div}\,b\right)u^2 = fu.$$

For justifying the previous identity, we used the Leibnitz rule

$$\partial_i(H\psi) = \psi\,\partial_i H + H\,\partial_i\psi, \tag{3.13}$$

valid for $\psi \in C^\infty$ and H any distribution. Then, integrating over $x \in \mathbb{R}^d$ we get

$$\frac{d}{dt}\int_{\mathbb{R}^d} u(t,x)^2 dx = 2\int_{\mathbb{R}^d} fu\,dx - 2\int_{\mathbb{R}^d}\left(c + \frac{1}{2}\text{div}\,b\right)u^2 dx$$

in the sense of distributions in $[0, T]$. Therefore, we get for a.e. $t \in [0, T]$

$$\left|\frac{d}{dt}\int_{\mathbb{R}^d} u(t,x)^2 dx\right|$$
$$\leq 2\|f(t,\cdot)\|_{L^2_x}\|u(t,\cdot)\|_{L^2_x} + 2\left\|\left(c + \frac{1}{2}\text{div}\,b\right)(t,\cdot)\right\|_{L^\infty_x}\|u(t,\cdot)\|_{L^2_x}^2.$$

This differential inequality can be easily integrated, obtaining

$$\|u(t,\cdot)\|_{L^2_x} \leq \|u(0,\cdot)\|_{L^2_x}\exp\left(\int_0^t\left\|\left(c + \frac{1}{2}\text{div}\,b\right)(s,\cdot)\right\|_{L^\infty_x} ds\right)$$
$$+ \int_0^t\|f(s,\cdot)\|_{L^2_x}\exp\left(\int_s^t\left\|\left(c + \frac{1}{2}\text{div}\,b\right)(\tau,\cdot)\right\|_{L^\infty_x} d\tau\right) ds,$$

which clearly implies (3.12). In the general case of $u \in \mathcal{F}^0$, we can find approximations u_n smooth with compact support in x such that $\|u_n - u\|_{\mathcal{F}} \to 0$, and we obtain the estimate (3.12) at the limit.

STEP 2. THE OPERATOR A^0. As in the proof of Theorem 3.1.1, we consider the linear operator

$$A^0: \begin{matrix} \mathcal{F}^0 & \to & L^2(\mathbb{R}^d) \times L^2([0,T] \times \mathbb{R}^d), \\ u & \mapsto & (u(0,\cdot),\ \partial_t u + \text{div}\,(bu) + cu). \end{matrix}$$

Since we can estimate

$$
\begin{aligned}
\|A^0 u\|_{L^2_x \times L^2_{t,x}} &= \|u(0, \cdot)\|_{L^2_x} + \|\partial_t u + \operatorname{div}(bu) + cu\|_{L^2_{t,x}} \\
&\leq \|u\|_{B_t(L^2_x)} + \|\partial_t u + \operatorname{div}(bu)\|_{L^2_{t,x}} + \|c\|_{L^\infty_{t,x}} \sqrt{T} \|u\|_{B_t(L^2_x)} \\
&\leq \left(1 + \|c\|_{L^\infty_{t,x}} \sqrt{T}\right) \|u\|_{\mathcal{F}},
\end{aligned}
$$

we deduce that A^0 is a bounded operator. Next, the energy estimate established in the first step gives that for any $u \in \mathcal{F}^0$,

$$
\|u\|_{B_t(L^2_x)} \leq \exp\left(T \left\|c + \frac{1}{2}\operatorname{div} b\right\|_{L^\infty_{t,x}}\right) \max(1, \sqrt{T}) \|A^0 u\|_{L^2_x \times L^2_{t,x}}.
$$

But we have

$$
\|\partial_t u + \operatorname{div}(bu)\|_{L^2_{t,x}} \leq \|\partial_t u + \operatorname{div}(bu) + cu\|_{L^2_{t,x}} + \|c\|_{L^\infty_{t,x}} \sqrt{T} \|u\|_{B_t(L^2_x)},
$$

and we conclude that

$$
\|u\|_{\mathcal{F}} \leq C \|A^0 u\|_{L^2_x \times L^2_{t,x}}, \qquad u \in \mathcal{F}^0. \tag{3.14}
$$

This means that A^0 is injective and with closed image. Notice that the injectivity of A^0 is equivalent to the fact that the only solution $u \in \mathcal{F}^0$ to

$$
\begin{cases}
\partial_t u + \operatorname{div}(bu) + cu = 0 \\
u(0, \cdot) = 0
\end{cases}
$$

is $u \equiv 0$.

STEP 3. PROOF OF THE EQUIVALENCE OF THE TWO STATEMENTS.
Since, by Step 2, A^0 is injective with closed image, we can apply Lemma 3.2.3(ii) to get the surjectivity of the adjoint operator $(A^0)^* : L^2(\mathbb{R}^d) \times L^2([0, T] \times \mathbb{R}^d) \to (\mathcal{F}^0)^*$. We recall that the adjoint operator is characterized by the condition

$$
\begin{aligned}
\langle (A^0)^*(v_0, v), u \rangle &= \langle (v_0, v), A^0 u \rangle \\
&= \int_{\mathbb{R}^d} v_0 u(0, \cdot) \, dx \\
&\quad + \int_{[0,T] \times \mathbb{R}^d} v\big(\partial_t u + \operatorname{div}(bu) + cu\big) \, dt dx,
\end{aligned} \tag{3.15}
$$

for $(v_0, v) \in L^2(\mathbb{R}^d) \times L^2([0, T] \times \mathbb{R}^d)$ and $u \in \mathcal{F}^0$. Since $(A^0)^*$ is surjective, in particular it has closed image. Therefore, applying Lemma 3.2.3(i) we get the equivalence between surjectivity of A^0 and injectivity of $(A^0)^*$.

It is clear that the surjectivity of the operator A^0 is equivalent to the existence of solutions in $\mathcal{F}^0$ (statement (i)). Therefore, it remains only to characterize the injectivity of $(A^0)^*$. Recalling the definition of $\mathcal{F}^0$ as the closure of the set of smooth functions with compact support in x, and recalling the characterization of the adjoint operator given in (3.15), we obtain that the injectivity of $(A^0)^*$ is equivalent to the following implication:

$$\left\{ \begin{array}{c} \displaystyle\int_{[0,T]\times\mathbb{R}^d} \big(\partial_t u + \text{div}\,(bu) + cu\big)v\,dt\,dx + \int_{\mathbb{R}^d} u(0,\cdot)v_0\,dx = 0 \\[2mm] \forall u \in C^\infty([0,T]\times\mathbb{R}^d) \text{ with compact support in } x \end{array} \right\} \qquad (3.16)$$

$$\implies \left\{ \begin{array}{c} v_0 = 0 \\ v = 0 \end{array} \right\}.$$

We argue as in Step 2 of the proof of Theorem 3.1.1, and test the integral condition with smooth functions of the form $u(t,x) = \chi(t/\epsilon)\tilde{u}(t,x)$ (using the same notation as in the proof of Theorem 3.1.1). Then, we obtain that the following two properties are equivalent for any given $v_0 \in L^2(\mathbb{R}^d)$ and $v \in L^2([0,T]\times\mathbb{R}^d)$:

1. For every $u \in C^\infty([0,T]\times\mathbb{R}^d)$ with compact support in x we have

$$\int_{[0,T]\times\mathbb{R}^d} \big(\partial_t u + \text{div}\,(bu) + cu\big)v\,dt\,dx + \int_{\mathbb{R}^d} u(0,\cdot)v_0\,dx = 0.$$

2. $v \in C([0,T]; L^2(\mathbb{R}^d) - w)$, $v_0 = v(0,\cdot)$, and v is a weak solution of the backward dual Cauchy problem

$$\left\{ \begin{array}{l} \partial_t v + b \cdot \nabla v - cv = 0 \\ v(T,\cdot) = 0. \end{array} \right.$$

Therefore we deduce that the implication (3.16) is equivalent to the uniqueness of weak solutions in $C([0,T]; L^2(\mathbb{R}^d) - w)$ of the backward dual Cauchy problem, i.e., statement (ii). □

Remark 3.2.4 (Time inversion). By reversing the direction of time, we see that there is existence for the backward Cauchy problem in $\mathcal{F}^0$ if and only if there is uniqueness for weak solutions to the forward dual Cauchy problem.

Remark 3.2.5. (Approximation by smooth functions and renormalization). Solutions in $\mathcal{F}^0$ lie in $C([0,T], L^2(\mathbb{R}^d) - s)$ and are renormalized: this can be seen as in the proof of the implication (ii) $\Rightarrow$ (iii) of

Theorem 3.1.1, using the density of smooth functions in $\mathcal{F}^0$. Conversely, it is possible that some renormalized solutions do not belong to $\mathcal{F}^0$. This can be seen by noticing that one can have several renormalized solutions to the same Cauchy problem (see an example in [84]), while there is always uniqueness in $\mathcal{F}^0$. Another difference between the criterion of approximation by smooth functions and the renormalization property is that $\mathcal{F}^0$ is a vector space, while in general renormalized solutions are not a vector space.

Remark 3.2.6 (Depauw's example again).   We notice that forward uniqueness and backward uniqueness of weak solutions are really distinct properties: the example described in Remark 3.1.5 shows how to construct bounded divergence-free vector fields with backward uniqueness, but not forward uniqueness, and vice versa.

3.3. Nearly incompressible vector fields

We recall (see Definition 2.8.1) that a vector field $b \in L^\infty([0, T] \times \mathbb{R}^d; \mathbb{R}^d)$ is *nearly incompressible* if there exists a function $\rho \in C([0, T]; L^\infty(\mathbb{R}^d) - w^*)$, with $0 < C^{-1} \leq \rho \leq C < \infty$ for some constant $C > 0$, such that the identity

$$\partial_t \rho + \mathrm{div}\,(b\rho) = 0 \qquad (3.17)$$

holds in the sense of distributions in $[0, T] \times \mathbb{R}^d$.

Theorem 3.3.1. *Let $b \in L^\infty([0, T] \times \mathbb{R}^d; \mathbb{R}^d)$ be a nearly incompressible vector field, and fix an associated function $\rho \in C([0, T]; L^\infty(\mathbb{R}^d) - w^*)$ as in Definition 2.8.1. We define the Banach space $\mathcal{F}$ and its norm $\|\cdot\|_{\mathcal{F}}$ as in (3.2)–(3.3). Let $\mathcal{F}^1 \subset \mathcal{F}$ be the closure of*

$$\{\rho\varphi \,:\, \varphi \in C^\infty([0, T] \times \mathbb{R}^d) \text{ with compact support in } x\}$$

with respect to $\|\cdot\|_{\mathcal{F}}$. Then the following statements are equivalent:

(i) *For every $u^0 \in L^2$ and every $f \in L^2$ there exists a solution $u \in \mathcal{F}^1$ to the Cauchy problem*

$$\begin{cases} \partial_t u + \mathrm{div}\,(bu) = f \\ u(0, \cdot) = u^0, \end{cases} \qquad u \in \mathcal{F}^1.$$

(ii) *There is uniqueness for weak solutions in $C([0, T]; L^2(\mathbb{R}^d) - w)$ for the backward dual Cauchy problem starting from T; i.e., the only function ρv belonging to $C([0, T]; L^2(\mathbb{R}^d) - w)$ which solves*

$$\begin{cases} \partial_t(\rho v) + \mathrm{div}\,(b\rho v) = 0 \\ \rho(T, \cdot)v(T, \cdot) = 0 \end{cases}$$

is $\rho v \equiv 0$.

Remark 3.3.2. In this context, the equation $\partial_t(\rho v) + \operatorname{div}(b\rho v) = 0$ is dual to the equation $\partial_t u + \operatorname{div}(bu) = 0$, since we can write (formally, since it is not possible to give a meaning to the product $b \cdot \nabla v$ without a condition of absolute continuity of $\operatorname{div} b$)

$$\partial_t(\rho v) + \operatorname{div}(b\rho v) = \rho(\partial_t v + b \cdot \nabla v).$$

Proof of Theorem 3.3.1. The proof is very close to that of Theorem 3.2.1; thus we shall sometimes omit the technical details.

STEP 1. AN ENERGY ESTIMATE IN $\mathcal{F}^1$. We preliminarily prove that for every $u \in \mathcal{F}^1$ the following estimate holds (C is the constant in (2.34)):

$$\|u\|_{B_t(L_x^2)} \leq C\|u(0, \cdot)\|_{L_x^2} + C\sqrt{T}\|\partial_t u + \operatorname{div}(bu)\|_{L_{t,x}^2}. \qquad (3.18)$$

Fix a smooth function φ with compact support in $\mathbb{R}^d$, and define

$$f = \partial_t(\rho\varphi) + \operatorname{div}(b\rho\varphi) = \rho(\partial_t\varphi + b \cdot \nabla\varphi)$$

(use the Leibniz rule (3.13) and formula (3.17)). We deduce with the same argument that

$$2\varphi f = \rho(\partial_t\varphi^2 + b \cdot \nabla\varphi^2) = \partial_t(\rho\varphi^2) + \operatorname{div}(b\rho\varphi^2).$$

Thus, we get the following estimate in the sense of distributions in $[0, T]$:

$$
\begin{aligned}
\frac{d}{dt}\int_{\mathbb{R}^d}\rho(t, x)\varphi(t, x)^2\,dx &= 2\int_{\mathbb{R}^d}\varphi(t, x)f(t, x)\,dx \\
&\leq 2\|f(t, \cdot)\|_{L_x^2}\|\varphi(t, \cdot)\|_{L_x^2} \\
&\leq 2\sqrt{C}\|f(t, \cdot)\|_{L_x^2}\left[\int_{\mathbb{R}^d}\rho(t, x)\varphi(t, x)^2\,dx\right]^{1/2}.
\end{aligned}
$$

By integration with respect to time this implies

$$
\left[\int_{\mathbb{R}^d}\rho(t, x)\varphi(t, x)^2\,dx\right]^{1/2} \leq \left[\int_{\mathbb{R}^d}\rho(0, x)\varphi(0, x)^2\,dx\right]^{1/2} \\
+ \sqrt{C}\int_0^t\|f(s, \cdot)\|_{L_x^2}\,ds.
$$

Using (2.34) we deduce

$$\frac{1}{\sqrt{C}}\|\rho(t, \cdot)\varphi(t, \cdot)\|_{L_x^2} \leq \sqrt{C}\|\rho(0, \cdot)\varphi(0, \cdot)\|_{L_x^2} + \sqrt{C}\int_0^t\|f(s, \cdot)\|_{L_x^2}\,ds,$$

and thus

$$\|\rho(t, \cdot)\varphi(t, \cdot)\|_{L_x^2} \leq C\|\rho(0, \cdot)\varphi(0, \cdot)\|_{L_x^2} + C\sqrt{T}\|\partial_t(\rho\varphi) \\ + \operatorname{div}(b\rho\varphi)\|_{L_{t,x}^2}. \tag{3.19}$$

But by definition of $\mathcal{F}^1$, the validity of (3.19) for every smooth function φ with compact support in x implies the validity of (3.18) for every function $u \in \mathcal{F}^1$.

STEP 2. THE OPERATOR A^1. We define the linear operator

$$A^1 : \begin{array}{rcl} \mathcal{F}^1 & \to & L^2(\mathbb{R}^d) \times L^2([0, T] \times \mathbb{R}^d), \\ \\ u & \mapsto & \big(u(0, \cdot),\ \partial_t u + \operatorname{div}(bu)\big). \end{array}$$

It is immediate to see that the operator A^1 is bounded. Using the energy estimate (3.18) it is also immediate to check that $\|u\|_{\mathcal{F}} \leq \tilde{C}\|A^1 u\|$, and therefore that A^1 is injective with closed image. Applying Lemma 3.2.3(ii) we obtain that the adjoint operator

$$(A^1)^* : L^2(\mathbb{R}^d) \times L^2([0, T] \times \mathbb{R}^d) \to (\mathcal{F}^1)^*$$

is surjective. The adjoint operator is characterized by the identity

$$\langle (A^1)^*(v_0, v), u \rangle = \langle (v_0, v), A^1 u \rangle$$

$$= \int_{\mathbb{R}^d} v_0 u(0, \cdot)\, dx \tag{3.20}$$

$$+ \int_{[0,T]\times\mathbb{R}^d} v\big(\partial_t u + \operatorname{div}(bu)\big)\, dt dx \tag{3.21}$$

for $(v_0, v) \in L^2(\mathbb{R}^d) \times L^2([0, T] \times \mathbb{R}^d)$ and $u \in \mathcal{F}^1$.

STEP 3. PROOF OF THE EQUIVALENCE OF THE TWO STATEMENTS. Statement (i) (existence of solutions in $\mathcal{F}^1$) is the surjectivity of the operator A^1, which is equivalent (applying Lemma 3.2.3(i) and using the surjectivity of $(A^1)^*$ proved in Step 2) to the injectivity of $(A^1)^*$. But recalling the characterization (3.21) and the definition of the space $\mathcal{F}^1$, we see that the injectivity of $(A^1)^*$ is equivalent to the following implication for $v_0 \in L^2(\mathbb{R}^d)$ and $v \in L^2([0, T] \times \mathbb{R}^d)$:

$$\left[\begin{array}{l} \int_{[0,T]\times\mathbb{R}^d} (\partial_t(\rho\varphi) + \operatorname{div}(b\rho\varphi))v\, dt dx + \int_{\mathbb{R}^d} \rho(0, \cdot)\varphi(0, \cdot)v_0\, dx = 0 \\ \\ \quad \forall \varphi \in C^\infty([0, T] \times \mathbb{R}^d) \text{ with compact support in } x \end{array} \right]$$

$$\implies \left\{ \begin{array}{l} v_0 = 0 \\ v = 0 \end{array} \right\}.$$

$$\tag{3.22}$$

Arguing as in Step 3 of the proof of Theorem 3.2.1 we obtain that the following two properties are equivalent:

- For every $\varphi \in C^{\infty}([0, T] \times \mathbb{R}^d)$ with compact support in x we have

$$\int_{[0,T]\times\mathbb{R}^d} \big(\partial_t(\rho\varphi) + \mathrm{div}\,(b\rho\varphi)\big) v \, dt dx + \int_{\mathbb{R}^d} \rho(0, \cdot)\varphi(0, \cdot)v_0 \, dx = 0.$$

- $\rho v \in C([0, T]; L^2(\mathbb{R}^d) - w)$, $\rho(0, \cdot)v_0 = \rho(0, \cdot)v(0, \cdot)$, and ρv is a weak solution of the backward dual Cauchy problem

$$\begin{cases} \partial_t(\rho v) + \mathrm{div}\,(b\rho v) = 0 \\ \rho(T, \cdot)v(T, \cdot) = 0. \end{cases}$$

Then we deduce that implication (3.22) is equivalent to statement (ii), and this concludes the proof of the theorem. $\qquad\square$

Chapter 4
Well-posedness in the two-dimensional case

In this chapter we describe some well-posedness results that are available in the two-dimensional case. Due to the special structure of the problem, which admits a Hamiltonian function conserved (at least formally) by the flow, the assumptions needed for the uniqueness are dramatically weaker than those of Chapter 2. We start in the first section by presenting some standard considerations and collecting some results available in the literature; we also mention a first result obtained with Colombini and Rauch [57] which goes beyond the divergence-free assumption. The rest of the chapter is devoted to the presentation of a work in progress in collaboration with Alberti and Bianchini [5], in which sharp well-posedness results in the two-dimensional case are obtained. We present here just the basic case of a bounded divergence-free vector field, while some variations are possible. The uniqueness holds under an additional assumption: we must require the weak Sard property (4.11), which turns out to be necessary, in view of the counterexamples contained in [5].

4.1. Bounded planar divergence-free vector fields

Let us consider an autonomous vector field $b \in L^\infty(\mathbb{R}^2; \mathbb{R}^2)$ in the plane such that $\text{div}\, b = 0$. It is well-known that in this situation it is possible to find a Hamiltonian function $H \in \text{Lip}(\mathbb{R}^2)$ such that

$$b(x) = \nabla^\perp H(x) = \left(-\frac{\partial H(x)}{\partial x_2}, \frac{\partial H(x)}{\partial x_1} \right) \quad \text{for } \mathscr{L}^2\text{-a.e. } x \in \mathbb{R}^2.$$

(4.1)

The starting point for all the two-dimensional well-posedness results is the heuristic remark that the value of the Hamiltonian is constant on the trajectories. Indeed, if $\dot{\gamma}(t) = b(\gamma(t))$, then we can compute

$$\frac{d}{dt}H(\gamma(t)) = \nabla H(\gamma(t)) \cdot \dot{\gamma}(t) = \nabla H(\gamma(t)) \cdot b(\gamma(t))$$

$$= \nabla H(\gamma(t)) \cdot \nabla^\perp H(\gamma(t)) = 0.$$

This means that the trajectories "follow" the level sets of the Hamiltonian. Heuristically, one can try to implement the following strategy:

(a) Localize the equation to each level set, thanks to the fact that the level sets are invariant under the action of the flow;
(b) Understand the structure of the level sets, trying to prove that generically they are "one-dimensional sets";
(c) See the equation on each level set as a one-dimensional problem and show uniqueness for it;
(d) Deduce uniqueness for the problem in $\mathbb{R}^2$ from the uniqueness of all the problems on the level sets.

Since we can hope for uniqueness on the level sets under quite general hypotheses, the reduced equation being one-dimensional, we expect stronger well-posedness results in this case: it is natural to imagine that no regularity of b (in terms of weak derivatives) would be needed.

We first indicate the essential literature on this subject. Previous results by Bouchut and Desvillettes [38], Hauray [100] and Colombini and Lerner ([58, 60]) show that uniqueness holds for the transport equation relative to an autonomous bounded divergence-free vector field, under the following additional condition on the local direction of the vector field: there exists an open set $\Omega \subset \mathbb{R}^2$ such that $\mathscr{H}^1(\mathbb{R}^2 \setminus \Omega) = 0$ and for every $x \in \Omega$ the following holds:

$$\text{there exist } \xi \in \mathbb{S}^1, \alpha > 0 \text{ and } \epsilon > 0 \text{ such that,} \tag{4.2}$$
$$\text{for } \mathscr{L}^2\text{-a.e. } y \in B_\epsilon(x), \text{ we have } b(y) \cdot \xi \geq \alpha.$$

The validity of this condition permits a local change of variable, which straightens the level sets of the Hamiltonian, thus reducing the equation to a one-dimensional problem (the second spatial variable appears as a parameter in the equation after the change of variable). A first extension to the non-divergence-free case is due to Colombini and Rauch [63]: they are able to show that the uniqueness holds in the case of autonomous bounded vector fields with bounded divergence for which there exists a positive Lipschitz function θ, bounded and bounded away from zero, such that

$$\text{div}\,(\theta b) = 0. \tag{4.3}$$

In [63] it is also conjectured that this hypothesis on the existence of the function θ could be removed. In a subsequent paper in collaboration with Colombini and Rauch [57] we show that this is the case: in fact we are able to prove the following result.

Theorem 4.1.1. *Assume that $b \in L^\infty(\mathbb{R}^2; \mathbb{R}^2)$ and that $\operatorname{div} b \in L^\infty(\mathbb{R}^2)$. Assume that there exists an open set $\Omega \subset \mathbb{R}^2$ such that $\mathcal{H}^1(\mathbb{R}^2 \setminus \Omega) = 0$ and that the condition in (4.2) holds for every $x \in \Omega$. Then we have uniqueness in $L^\infty([0, T] \times \mathbb{R}^2)$ for the Cauchy problem for the transport equation.*

However, the meaning of condition (4.2) is not completely clear: while in the stationary problem it just expresses the fact the surface on which we consider the initial data is noncharacteristic, in the time-space problem it is a kind of local regularity of the direction of b. In particular, condition (4.2) prevents the existence of "too many" zeros of the vector field.

In the rest of this chapter we present a more recent result, still in progress, in collaboration with Alberti and Bianchini [5]. In this paper the strategy is a bit different: we do not perform a local change of variable according to the Hamiltonian, but we rather split the equation on the level sets of the Hamiltonian, using the coarea formula. Then we would like to look at the equation level set by level set. It turns out that, where $\nabla H \neq 0$, the level sets are in fact nice rectifiable curves, and this will allow to consider the PDE in the parametrization. The interesting point is that, in order to separate the evolution in $\{\nabla H = 0\}$ from the evolution in $\{\nabla H \neq 0\}$, we need a condition which is reminiscent of (4.2), in the sense that it regards again the "amount of the critical points of H". This is precisely the weak Sard property in (4.11). However we notice that condition (4.11) is much weaker than the previous one; moreover, two examples (for which we refer to [5]) indicate that the weak Sard property is necessary in order to obtain uniqueness. We present here a detailed account of our proof in the basic case of a bounded divergence-free vector field: we remark that various generalizations and extensions are possible (we refer again to [5]).

4.2. Splitting on the level sets of the Hamiltonian

We will be concerned with the Cauchy problem for the transport equation

$$\begin{cases} \partial_t u + b \cdot \nabla u = 0 \\ u(0, \cdot) = \bar{u} \end{cases} \quad \text{in } \mathcal{D}'([0, T] \times \mathbb{R}^2), \qquad (4.4)$$

where $\bar{u} \in L^\infty(\mathbb{R}^2)$. We assume that $b \in L^\infty(\mathbb{R}^2; \mathbb{R}^2)$ has compact support and that $\operatorname{div} b = 0$. Recall that this implies the existence of a Hamiltonian function $H \in \operatorname{Lip}_c(\mathbb{R}^2)$ which satisfies (4.1); we denote by $\operatorname{Lip}_c(X)$ the space of Lipschitz functions with compact support defined on X. We remark that the compactness of the support has been assumed in order to simplify the classification of the level sets in Theorem 4.4.1

and to avoid localizations in the equation, but it is not really essential for our problem.

We consider the weak formulation of (4.4): $u(t, x) \in L^\infty([0, T] \times \mathbb{R}^2)$ is a weak solution of (4.4) if for every $\varphi(t, x) \in \text{Lip}_c([0, T[\times \mathbb{R}^2)$ we have

$$\int_0^T \int_{\mathbb{R}^2} u(\partial_t \varphi + b \cdot \nabla \varphi) \, dx dt = - \int_{\mathbb{R}^2} \bar{u} \varphi(0, \cdot) \, dx. \quad (4.5)$$

We now recall a particular case of the coarea formula (A.3). For every $h \in \mathbb{R}$ we denote by E_h the level sets

$$E_h = \{x \in \mathbb{R}^2 : H(x) = h\}.$$

Then, for $H \in \text{Lip}_c(\mathbb{R}^2)$ and for every function $\phi \in L^1(\mathbb{R}^2)$, we have

$$\int_{\mathbb{R}^2 \cap \{\nabla H \neq 0\}} \phi \, dx = \int_{\mathbb{R}} \left[\int_{E_h} \frac{\phi}{|\nabla H|} \, d\mathcal{H}^1 \right] dh. \quad (4.6)$$

Using (4.6) in (4.5) and recalling that $b = \nabla^\perp H$ we obtain

$$0 = \int_0^T \int_{\{\nabla H = 0\}} u \partial_t \varphi \, dx dt + \int_{\{\nabla H = 0\}} \bar{u} \varphi(0, \cdot) \, dx$$
$$+ \int_0^T \int_{\{\nabla H \neq 0\}} u(\partial_t \varphi + b \cdot \nabla \varphi) \, dx dt + \int_{\{\nabla H \neq 0\}} \bar{u} \varphi(0, \cdot) \, dx$$
$$= \int_0^T \int_{\{\nabla H = 0\}} u \partial_t \varphi \, dx dt + \int_{\{\nabla H = 0\}} \bar{u} \varphi(0, \cdot) \, dx \quad (4.7)$$
$$+ \int_0^T \int_{\mathbb{R}} \left[\int_{E_h} \frac{u}{|\nabla H|} (\partial_t \varphi + b \cdot \nabla \varphi) \, d\mathcal{H}^1 \right] dh dt$$
$$+ \int_{\mathbb{R}} \left[\int_{E_h} \frac{\bar{u}}{|\nabla H|} \varphi(0, \cdot) \, d\mathcal{H}^1 \right] dh.$$

The following lemma will allow the selection of the level sets of the Hamiltonian. We recall that we denote by $H_\# \mathcal{L}^2$ the push-forward of the Lebesgue measure on $\mathbb{R}^2$ via the function H, defined according to (A.1).

Lemma 4.2.1. *If $u(t, x)$ is a weak solution of (4.4) and $\eta(h) \in L^1(\mathbb{R}, H_\# \mathcal{L}^2)$, then $u(t, x)\eta(H(x))$ is a weak solution of (4.4).*

Proof. Considering the weak formulation (4.5) with test function

$$\psi(t, x) = \varphi(t, x)\eta(H(x))$$

we deduce the validity of the lemma for any Lipschitz function η. The thesis for every $\eta \in L^1(\mathbb{R}, H_\# \mathcal{L}^2)$ follows from an approximation procedure, since no derivatives of η are involved in the weak formulation. $\square$

We now introduce some notation that will be used in the rest of this chapter. We consider the measure λ_φ defined by

$$\lambda_\varphi = H_\# \left(\left(\int_0^T u \partial_t \varphi \, dt + \bar{u}\varphi(0, \cdot) \right) \mathcal{L}^2 \llcorner \{\nabla H = 0\} \right).$$

It is readily checked that $\lambda_\varphi \ll H_\#(\mathcal{L}^2 \llcorner \{\nabla H = 0\})$. We denote by $\lambda_\varphi(h)$ the density of λ_φ with respect to $H_\#(\mathcal{L}^2 \llcorner \{\nabla H = 0\})$, i.e.

$$\lambda_\varphi = \lambda_\varphi(h) H_\#(\mathcal{L}^2 \llcorner \{\nabla H = 0\}).$$

Moreover, for every $\eta \in L^1(\mathbb{R}, H_\# \mathcal{L}^2)$, we have

$$\lambda_{\eta(H)\varphi} = \eta(h)\lambda_\varphi(h) H_\#(\mathcal{L}^2 \llcorner \{\nabla H = 0\}).$$

We perform the decomposition of $H_\#(\mathcal{L}^2 \llcorner \{\nabla H = 0\})$ into the absolutely continuous and the singular (with respect to $\mathcal{L}^1$) parts:

$$H_\#(\mathcal{L}^2 \llcorner \{\nabla H = 0\}) = \left[H_\#(\mathcal{L}^2 \llcorner \{\nabla H = 0\}) \right]^a(h) \mathcal{L}^1$$
$$+ \left[H_\#(\mathcal{L}^2 \llcorner \{\nabla H = 0\}) \right]^s.$$

Going back to (4.7), using Lemma 4.2.1 and the notation introduced we obtain, for every $\eta \in L^1(\mathbb{R}, H_\# \mathcal{L}^2)$,

$$\int_\mathbb{R} \eta(h)\lambda_\varphi(h) d\left[H_\#(\mathcal{L}^2 \llcorner \{\nabla H = 0\}) \right](h)$$
$$+ \int_0^T \int_\mathbb{R} \eta(h) \left[\int_{E_h} \frac{u}{|\nabla H|} (\partial_t \varphi + b \cdot \nabla \varphi) \, d\mathcal{H}^1 \right] dh dt \qquad (4.8)$$
$$+ \int_\mathbb{R} \eta(h) \left[\int_{E_h} \frac{\bar{u}}{|\nabla H|} \varphi(0, \cdot) \, d\mathcal{H}^1 \right] dh = 0.$$

The arbitrariness of the function $\eta \in L^1(\mathbb{R}, H_\# \mathcal{L}^2)$ in (4.8) then gives the following:

(i) for $\mathcal{L}^1$-a.e. $h \in \mathbb{R}$ we have

$$\lambda_\varphi(h) \left[H_\#(\mathcal{L}^2 \llcorner \{\nabla H = 0\}) \right]^a(h)$$
$$+ \int_0^T \int_{E_h} \frac{u}{|\nabla H|} (\partial_t \varphi + b \cdot \nabla \varphi) \, d\mathcal{H}^1 dt \qquad (4.9)$$
$$+ \int_{E_h} \frac{\bar{u}}{|\nabla H|} \varphi(0, \cdot) \, d\mathcal{H}^1 = 0;$$

(ii) for $\left[H_\#(\mathcal{L}^2 \llcorner \{\nabla H = 0\}) \right]^s$-a.e. $h \in \mathbb{R}$ we have

$$\lambda_\varphi(h) = 0. \qquad (4.10)$$

4.3. The weak Sard property

We see from equation (4.9) that the dynamics in $\{\nabla H \neq 0\}$ and in $\{\nabla H = 0\}$ could be coupled: this can actually happen, as shown in the examples constructed in [5]. This means that we can have interactions between the areas in which the velocity is zero and the ones in which it is nonzero. In order to separate the two dynamics we need the following *weak Sard property* of the Hamiltonian.

Definition 4.3.1. We say that $H \in \mathrm{Lip}_c(\mathbb{R}^2)$ satisfies the *weak Sard property* if

$$H_\#\big(\mathscr{L}^2 \llcorner \{\nabla H = 0\}\big) \perp \mathscr{L}^1. \qquad (4.11)$$

Using the notation introduced in the previous section this means that

$$\big[H_\#\big(\mathscr{L}^2 \llcorner \{\nabla H = 0\}\big)\big]^a(h) = 0 \qquad \text{for } \mathscr{L}^1\text{-a.e. } h \in \mathbb{R}.$$

The connection with the classical Sard theorem (see for instance [92, Theorem 3.4.3]) is evident: here we are requiring that the "image" (via the push-forward through H) of the Lebesgue measure $\mathscr{L}^2$ restricted to the set of the critical points $\{\nabla H = 0\}$ is "not seen" by the Lebesgue measure $\mathscr{L}^1$ in the codomain.

Assuming the weak Sard property we can separate the two dynamics, hence from equations (4.9) and (4.10) we deduce the following result.

Theorem 4.3.2. *Let $b \in L^\infty(\mathbb{R}^2; \mathbb{R}^2)$ with compact support and assume that $\mathrm{div}\, b = 0$. Let $H \in \mathrm{Lip}_c(\mathbb{R}^2)$ be as in (4.1) and assume that H satisfies the weak Sard property (4.11). Let $u \in L^\infty([0, T] \times \mathbb{R}^d)$ be a weak solution of (4.4). Then we have*

$$\int_0^T \int_{E_h} \frac{u}{|\nabla H|}(\partial_t \varphi + b \cdot \nabla \varphi)\, d\mathscr{H}^1 dt + \int_{E_h} \frac{\bar{u}}{|\nabla H|}\varphi(0, \cdot)\, d\mathscr{H}^1 = 0$$
$$(4.12)$$

for $\mathscr{L}^1$-a.e. $h \in \mathbb{R}$ and

$$\int_0^T \int_{\{\nabla H=0\}} u\partial_t \varphi\, dxdt + \int_{\{\nabla H=0\}} \bar{u}\varphi(0, \cdot)\, dx = 0. \qquad (4.13)$$

Notice that (4.13) gives $u(t, x) = \bar{u}$ for $\mathscr{L}^1 \otimes \mathscr{L}^2$-a.e. $(t, x) \in [0, T] \times \{\nabla H = 0\}$. This means that, thanks to the weak Sard property, the uniqueness for the Cauchy problem (4.4) is equivalent to the uniqueness for the "reduced problems" (4.12) on the level sets, for $\mathscr{L}^1$-a.e. $h \in \mathbb{R}$. The issue of the uniqueness on the level sets is discussed in Section 4.5.

We stress the fact that the weak Sard property (4.11) is not just a technical assumption that we need in order to implement our strategy: it is

really necessary for the well-posedness, as pointed out by the two counterexamples of [5], where nonuniqueness phenomena are shown in the
case of vector fields which do not satisfy (4.11).

4.4. Structure of the level sets

In this section we give a detailed description of the structure of the level
sets

$$E_h = \{x \in \mathbb{R}^2 \ : \ H(x) = h\}.$$

We first notice, by the continuity of H and by the assumption of compactness of the support, that for every $h \neq 0$ the set E_h is compact. Moreover,
applying the coarea formula (4.6) with $\phi \equiv 1$ we obtain

$$\int_{\mathbb{R}} \left[\int_{E_h} \frac{1}{|\nabla H|} \, d\mathcal{H}^1 \right] dh = \mathcal{L}^2(\{\nabla H \neq 0\}) < +\infty,$$

by the compactness of spt H. This implies that

$$\int_{E_h} \frac{1}{|\nabla H|} \, d\mathcal{H}^1 < +\infty \qquad \text{for } \mathcal{L}^1\text{-a.e. } h \in \mathbb{R}. \tag{4.14}$$

In particular, since $|\nabla H| \leq \|b\|_\infty$, this yields

$$\mathcal{H}^1(E_h) < +\infty \qquad \text{for } \mathcal{L}^1\text{-a.e. } h \in \mathbb{R}.$$

For every $h \in \mathbb{R}$, we denote by $\mathscr{C}_h$ the family of all the connected components C of E_h such that $\mathcal{H}^1(C) > 0$ (in fact, these are just the connected
components which contain more than one point).

We collect together in the following theorem all the results relative to
the classification of the level sets. For the proof we refer to [5]. See
Appendix A.1 for the notion of rectifiable set.

Theorem 4.4.1. *Let $H \in \text{Lip}_c(\mathbb{R}^2)$. For $\mathcal{L}^1$-a.e. $h \in \mathbb{R}$ the following
statements hold.*

(i) *E_h is $\mathcal{H}^1$-rectifiable and $\mathcal{H}^1(E_h) < +\infty$; the map H is differen-
tiable in x and $\nabla H \neq 0$ at $\mathcal{H}^1$-a.e. $x \in E_h$; the function $1/|\nabla H|$
belongs to $L^1(E_h, \mathcal{H}^1)$.*
(ii) *The family $\mathscr{C}_h$ is countable and $\mathcal{H}^1\left(E_h \setminus \cup_{C \in \mathscr{C}_h} C\right) = 0$.*
(iii) *Every $C \in \mathscr{C}_h$ is a closed simple curve. More precisely, it is possible
to find a Lipschitz injective parametrization $\gamma : [\alpha, \beta]^* \to C$ such
that*

$$\dot{\gamma}(s) = \nabla^\perp(\gamma(s)) \qquad \text{for } \mathcal{L}^1\text{-a.e. } s \in [\alpha, \beta]^*, \tag{4.15}$$

where we denote by $[\alpha, \beta]^*$ the quotient space consisting of the interval $[\alpha, \beta]$ with identified endpoints, endowed with the distance

$$\text{dist}_{[\alpha,\beta]^*}(x, y) = \min \left\{ |x - y|, (\beta - \alpha) - |x - y| \right\}.$$

We will also need the following topological lemma.

Lemma 4.4.2. *Let $h \in \mathbb{R}$ such that the conclusions of Theorem 4.4.1 hold. Then, for every $C \in \mathscr{C}_h$, there exists a decreasing sequence $\{U_n\}$ of bounded open sets in $\mathbb{R}^2$ such that $\partial U_n \cap E_h = \emptyset$ for every n and $E_h \cap (\cap_n U_n) = C$.*

4.5. Uniqueness on the level sets

We first show, using Lemma 4.4.2, that equation (4.12) can be separated into a family of equations on the connected components of each level set.

Proposition 4.5.1. *Let $b \in L^\infty(\mathbb{R}^2; \mathbb{R}^2)$ with compact support and assume that $\text{div } b = 0$. Let $H \in \text{Lip}_c(\mathbb{R}^2)$ be as in (4.1) and assume that H satisfies the weak Sard property (4.11). Let $u \in L^\infty([0, T] \times \mathbb{R}^d)$ be a weak solution of (4.4). Then, for every $C \in \mathscr{C}_h$, for $\mathscr{L}^1$-a.e. $h \in \mathbb{R}$, we have*

$$\int_0^T \int_C \frac{u}{|\nabla H|} (\partial_t \varphi + b \cdot \nabla \varphi) \, d\mathscr{H}^1 dt + \int_C \frac{\bar{u}}{|\nabla H|} \varphi(0, \cdot) \, d\mathscr{H}^1 = 0.$$
$$(4.16)$$

Proof. We fix $h \in \mathbb{R}$ such that the conclusions of Theorems 4.3.2 and 4.4.1 hold. We choose a sequence $\{U_n\}$ as in Lemma 4.4.2. Since ∂U_n and E_h are compact sets we have

$$\text{dist}(\partial U_n, E_h) = \epsilon_n > 0.$$

Thus we fix a standard convolution kernel ρ with $\text{spt } \rho \subset B_1(0)$ and for every n we set

$$\gamma_n(x) = \mathbf{1}_{U_n} * \rho_{\epsilon_n/4}(x).$$

We rewrite equation (4.12) with the test function $\varphi(t, x)\gamma_n(x)$. We have

$$0 = \int_0^T \int_{E_h} \frac{u}{|\nabla H|} \left(\partial_t \varphi \gamma_n + b \cdot \nabla (\varphi \gamma_n) \right) d\mathscr{H}^1 dt$$

$$+ \int_{E_h} \frac{\bar{u}}{|\nabla H|} \varphi(0, \cdot) \gamma_n \, d\mathscr{H}^1$$

$$= \int_0^T \int_{E_h \cap U_n} \frac{u}{|\nabla H|} (\partial_t \varphi + b \cdot \nabla \varphi) \, d\mathscr{H}^1 dt$$

$$+ \int_{E_h \cap U_n} \frac{\bar{u}}{|\nabla H|} \varphi(0, \cdot) \, d\mathscr{H}^1.$$

We now let $n \to \infty$ in the above equality. Recalling (4.14) and applying the Lebesgue dominated convergence theorem we eventually obtain (4.16). □

We are now in the position to formulate and prove our main result.

Theorem 4.5.2. *Let* $b \in L^\infty(\mathbb{R}^2; \mathbb{R}^2)$ *with compact support and assume that* $\operatorname{div} b = 0$. *Let* $H \in \operatorname{Lip}_c(\mathbb{R}^2)$ *be as in* (4.1) *and assume that* H *satisfies the weak Sard property* (4.11). *Then, for every initial data* $\bar{u} \in L^\infty(\mathbb{R}^2)$, *the Cauchy problem* (4.4) *has a unique solution* $u \in L^\infty([0, T] \times \mathbb{R}^2)$.

Proof. The existence is obtained by the same regularization technique used in Theorem 2.2.3. To show the uniqueness, by linearity it suffices to show that the only solution $u \in L^\infty([0, T] \times \mathbb{R}^2)$ with initial data $\bar{u} \equiv 0$ is $u \equiv 0$. Recalling the discussion at the end of Section 4.3 and the result of Proposition 4.5.1, it is enough to show that, for every $C \in \mathscr{C}_h$, for $\mathscr{L}^1$-a.e. $h \in \mathbb{R}$, the validity of

$$\int_0^T \int_C \frac{u}{|\nabla H|} (\partial_t \varphi + b \cdot \nabla \varphi) \, d\mathscr{H}^1 dt = 0$$

for every $\varphi(t, x) \in \operatorname{Lip}_c([0, T[\times \mathbb{R}^2)$ implies $u(t, x) = 0$ for $\mathscr{L}^1 \otimes \mathscr{H}^1$-a.e. $(t, x) \in [0, T] \times C$. We proceed in several steps.

STEP 1. PARAMETRIZATION OF C. We fix $h \in \mathbb{R}$ such that the conclusions of Theorem 4.4.1 and of Proposition 4.5.1 hold. Using the result in Theorem 4.4.1 (iii) we know that every $C \in \mathscr{C}_h$ is a closed simple curve and we choose a Lipschitz injective parametrization $\gamma : [\alpha, \beta]^* \to C$ which satisfies (4.15). Hence from (4.16) (with $\bar{u} \equiv 0$) we get

$$\int_0^T \int_\alpha^\beta u(t, \gamma(s)) \Big((\partial_t \varphi)(t, \gamma(s)) + b(\gamma(s)) \cdot (\nabla \varphi)(t, \gamma(s)) \Big) \, ds \, dt = 0$$
$$\tag{4.17}$$

for every $\varphi \in \operatorname{Lip}_c([0, T[\times \mathbb{R}^2)$. We set

$$\tilde{\varphi}(t, s) = \varphi(t, \gamma(s)). \tag{4.18}$$

Differentiating both sides of (4.18) with respect to s we get

$$\partial_s \tilde{\varphi}(t, s) = (\nabla \varphi)(t, \gamma(s)) \cdot \dot{\gamma}(s) = (\nabla \varphi)(t, \gamma(s)) \cdot (\nabla^\perp H)(\gamma(s)),$$

and this implies

$$b(\gamma(s)) \cdot (\nabla \varphi)(t, \gamma(s)) = (\nabla^\perp H)(\gamma(s)) \cdot (\nabla \varphi)(t, \gamma(s)) = \partial_s \tilde{\varphi}(t, s).$$
$$\tag{4.19}$$

Setting $\tilde{u}(t, s) = u(t, \gamma(s))$ and inserting (4.19) in (4.17) we obtain

$$\int_0^T \int_\alpha^\beta \tilde{u}\left(\partial_t \tilde{\varphi} + \partial_s \tilde{\varphi}\right) ds dt = 0 \qquad (4.20)$$

for every $\tilde{\varphi} : [0, T] \times [\alpha, \beta]^* \to \mathbb{R}$ of the form $\tilde{\varphi}(t, s) = \varphi(t, \gamma(s))$ for some $\varphi \in \mathrm{Lip}_c([0, T[\times\mathbb{R}^2)$.

STEP 2. TEST FUNCTIONS IN $[0, T] \times [\alpha, \beta]^*$. We notice that, up to now, we cannot see (4.20) as a distributional equation on $[0, T] \times [\alpha, \beta]^*$: indeed, we are allowed to use as test functions only the particular $\tilde{\varphi}$'s of the form above. However, the following lemma from [5] holds.

Lemma 4.5.3. *Every* $\psi \in \mathrm{Lip}_c([0, T[\times, [\alpha, \beta]^*)$ *can be approximated uniformly with a sequence of functions* $\{\tilde{\varphi}_n\}$ *of the form above and such that* $\mathrm{Lip}(\tilde{\varphi}_n)$ *is equi-bounded.*

This means that we can write (4.20) with $\tilde{\varphi} = \tilde{\varphi}_n$ for every n and passing to the limit we get

$$\int_0^T \int_\alpha^\beta \tilde{u}\left(\partial_t \psi + \partial_s \psi\right) ds dt = 0 \qquad (4.21)$$

for any $\psi \in \mathrm{Lip}_c([0, T[\times, [\alpha, \beta]^*)$. This is now a distributional equation on $[0, T] \times [\alpha, \beta]^*$.

STEP 3. UNIQUENESS ON C. Now it suffices to notice that (4.21) is the distributional form of the Cauchy problem

$$\begin{cases} \partial_t \tilde{u} + \partial_s \tilde{u} = 0 \\ \tilde{u}(0, \cdot) = 0. \end{cases} \qquad (4.22)$$

By the smooth theory for the transport equation of Section 1.4 we know that the only solution to this problem is $\tilde{u} \equiv 0$. From the definition of $\tilde{u}$ we see that this means that $u(t, x) = 0$ for $\mathscr{L}^1 \otimes \mathscr{H}^1$-a.e. $(t, x) \in [0, T] \times C$. We conclude the desired thesis. $\qquad \square$

We close this chapter by presenting a particular case in which the weak Sard property (4.11) is satisfied by the function $H \in \mathrm{Lip}_c(\mathbb{R}^2)$ associated to b as in (4.1) and thus the uniqueness result of Theorem 4.5.2 holds. See [5] for the proof.

Corollary 4.5.4. *Let* $b \in L^\infty(\mathbb{R}^2; \mathbb{R}^2)$ *with compact support and assume that* $\mathrm{div}\, b = 0$ *and that* b *is approximately differentiable* $\mathscr{L}^2$*-a.e. in* $\mathbb{R}^2$. *Then, for every initial data* $\bar{u} \in L^\infty(\mathbb{R}^2)$, *the Cauchy problem* (4.4) *has a unique solution* $u \in L^\infty([0, T] \times \mathbb{R}^2)$.

We observe that the approximate differentiability assumption on b in Corollary 4.5.4 is of "qualitative" type, in contrast with the weak regularity assumptions (for instance Sobolev or BV) in Chapter 2.

Chapter 5
Counterexamples to the well-posedness and related constructions

In this chapter we present some counterexamples, whose aim is to show the sharpness of the assumptions made in the various renormalization theorems. In the first section we illustrate an elegant construction by Depauw [82], which yields an explicit example of nonuniqueness for a vector field enjoying a regularity very close to the one needed in Ambrosio's theorem. This example has striking consequences on the possibility of building a theory for multi-dimensional systems of conservation laws based on transport equations: this is obtained in a joint work with De Lellis [65] and is briefly presented in Section 5.2. In Section 5.3 we make some comments on two counterexamples to the uniqueness (in fact at the ODE level) by DiPerna and Lions [84]. The last section is devoted to some examples by Colombini, Luo and Rauch [62], relative to the lack of propagation of classical regularities for vector fields which are less than Lipschitz.

5.1. Depauw's counterexample

Following some ideas of Aizenman ([2]), Depauw in [82] and Colombini, Luo and Rauch in [61] have recently given some counterexamples to the uniqueness for the transport equation with a bounded divergence-free vector field (see also [48] for a related construction).

These examples also show the sharpness of the result by Ambrosio in Theorem 2.6.1: for instance, the vector field constructed in [82] is "almost BV", in a sense that will be clear in a moment. The example of Depauw consists of a bounded planar divergence-free vector field $a(t, x)$ with two different bounded distributional solutions of

$$\begin{cases} \partial_t w + \operatorname{div}(aw) = 0 \\ w(0, \cdot) = 0 . \end{cases} \tag{5.1}$$

We want to show briefly the construction of this vector field. First of all we define $b : [-1/2, 1/2]^2 \to \mathbb{R}^2$ as

$$b(x_1, x_2) = \begin{cases} (0, 4x_1) & \text{if } 0 < |x_2| < |x_1| < 1/4 \\ (-4x_2, 0) & \text{if } 0 < |x_1| < |x_2| < 1/4 \\ 0 & \text{otherwise} \end{cases} \tag{5.2}$$

and we extend it periodically to $\mathbb{R}^2$ (see Figure 5.1). The field $a(t, x)$ is then given by

$$a(t, x) = \begin{cases} 0 & \text{if } t < 0 \text{ or } t > 1 \\ b(2^j x) & \text{if } t \in I_j = 2^{-j}(\frac{1}{2}, 1) \text{ for some } j \in \mathbb{N}, \end{cases}$$

and we also define $c(t, x) = a(1 - t, x)$. It follows immediately that a and c are bounded and divergence-free. Moreover, note that for $\mathscr{L}^1$-a.e. $t \in \mathbb{R}$ we have $a(t, \cdot) \in BV_{\text{loc}}(\mathbb{R}^2; \mathbb{R}^2)$, but a does not belong to $L^1([0, \alpha]; BV_{\text{loc}}(\mathbb{R}^2))$ for any $\alpha > 0$.

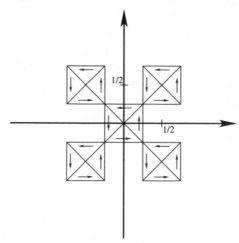

Figure 5.1. Depauw's vector field.

We want to describe the flow of c. First of all, we recall that for $0 \le t_0 < t_1 < 1$ we denote by $X^{(c)}(t_1, t_0, x)$ the solution at time t_1 of the problem

$$\begin{cases} \dot{\psi}(t) = c(t, \psi(t)) \\ \psi(t_0) = x. \end{cases}$$

Note that $X^{(c)}$ is well-defined since c is piecewise smooth on $[t_0, t_1] \times \mathbb{R}^2$. Next we let u_0 be the $\mathbb{Z}^2$-periodic function given by $u_0(x_1, x_2) = \text{sgn}(x_1 x_2)$ on the square $\left(-\frac{1}{2}, \frac{1}{2}\right)^2$. If we define $u_k(x) = u_0(2^k x)$, then $u_k\left(X^{(c)}\left(1 - 2^{-k}, 1 - 2^{-k-1}, x\right)\right) = u_{k+1}(x)$ (see Figure 5.2).

By the semigroup property of $X^{(c)}$ we conclude that $u_0(X^{(c)}(0, 1 - 2^{-k}, x)) = u_k(x)$. Therefore $u(t, x) = u_0(X^{(c)}(0, t, x))$ is a bounded distributional solution of

$$\begin{cases} \partial_t u + \operatorname{div}(uc) = 0 \\ u(0, \cdot) = u_0. \end{cases} \tag{5.3}$$

Note that $u(t, \cdot)$ converges weakly (but not strongly) to 0 as $t \uparrow 1$. Therefore w defined by $w(t, x) = u(1 - t, x)$ is a nontrivial weak solution of (5.1).

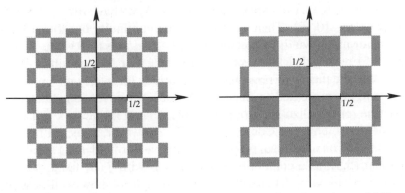

Figure 5.2. The effect of Depauw's vector field b acting for a time of $1/2$ on a chessboard of side $1/4$.

5.2. Oscillatory solutions to transport equations

The example of [82] presented in Section 5.1 has severe consequences on the possibility of building a theory for multi-dimensional systems of conservation laws based on transport equations. We recall the Keyfitz and Kranzer system

$$\begin{cases} \partial_t u + \sum_{j=1}^{d} \frac{\partial}{\partial x_j} (f^j(|u|)u) = 0 \\ u(0, \cdot) = \bar{u} \end{cases} \qquad u : [0, T] \times \mathbb{R}^d \to \mathbb{R}^k \tag{5.4}$$

already presented in the Introduction. For every $j = 1, \ldots, d$ the map $f^j : \mathbb{R}^+ \to \mathbb{R}$ is smooth. Notice the very special feature of this system: the nonlinearity only depends on the moduls of the solution. Even in the presence of this special "triangular" structure, Bressan [48] proved that the Cauchy problem (5.4) can be ill posed in L^∞, showing initial data $\bar{u}$ which generate renormalized entropy solutions with wild oscillations, whose weak limit is not a solution any more. We can decouple (5.4) in a scalar conservation law for the modulus $\rho = |u|$

$$\partial_t \rho + \operatorname{div}\left(f(\rho)\rho\right) = 0 \tag{5.5}$$

and a continuity equation for the angular part $\theta = u/|u|$

$$\partial_t(\rho\theta) + \operatorname{div}\left(f(\rho)\rho\theta\right) = 0. \tag{5.6}$$

The heuristic idea behind the well-posedness result for (5.4) in [15, 11] is to consider first Kružkov's entropy solutions of (5.5). If $|\bar{u}| \in BV(\mathbb{R}^d)$ then from the theory of entropy solutions [103] we have $\rho(t, \cdot) \in BV(\mathbb{R}^d)$ for every $t \in [0, T]$. Then we consider the continuity equation (5.6) for the vector unknown θ: the vector field $f(\rho)$ belongs to $BV([0, T] \times \mathbb{R}^d)$ and (5.5) means that this vector field is nearly incompressible. We observe that the theory of renormalized solutions for nearly incompressible vector fields presented in Section 2.8 does not cover immediately this case (the renormalization property in this general framework is presently an open problem: see Conjecture 6.7.2). However, some ad hoc arguments based on additional properties of the vector field $f(\rho)$ (namely the fact that (5.5), as a continuity equation in ρ with $f(\rho)$ fixed, possesses a solution enjoying BV regularity) are enough to show well-posedness for (5.6), in the class of the so-called renormalized entropy solutions (see [15, 11, 77] for the details).

It is therefore tempting to try to generalize this approach to more general systems of conservations laws. Another kind of prototype of multi-dimensional system consists of two continuity equations coupled through some nonlinearity:

$$\begin{cases} \partial_t u + \operatorname{div}\left(f(v)u\right) = 0 \\ \partial_t v + \operatorname{div}\left(g(u)v\right) = 0 \\ u(0, \cdot) = \bar{u} \\ v(0, \cdot) = \bar{v} \end{cases} \qquad \begin{array}{l} f, g : \mathbb{R} \to \mathbb{R}^d \\ u, v : [0, T] \times \mathbb{R}^d \to \mathbb{R}. \end{array} \tag{5.7}$$

The natural attempt would be the implementation of an iterative procedure. We first find u_1 and v_2 solutions to the two continuity equations with fixed vector fields:

$$\begin{cases} \partial_t u_1 + \operatorname{div}\left(f(\bar{v})u_1\right) = 0 \\ \partial_t v_1 + \operatorname{div}\left(g(\bar{u})v_1\right) = 0 \\ u_1(0, \cdot) = \bar{u} \\ v_2(0, \cdot) = \bar{v}. \end{cases}$$

We iterate this procedure, finding for every $k \in \mathbb{N}$ solutions u_k and v_k to the continuity equations in which the vector fields are fixed and depend on the solutions u_{k-1} and v_{k-1} found in the previous step:

$$\begin{cases} \partial_t u_k + \operatorname{div} \left(f(v_{k-1}) u_k \right) = 0 \\ \partial_t v_k + \operatorname{div} \left(g(u_{k-1}) v_k \right) = 0 \\ u_k(0, \cdot) = \bar{u} \\ v_k(0, \cdot) = \bar{v} . \end{cases}$$

Then one might expect convergence of the two sequences $\{u_k\}$ and $\{v_k\}$ to a solution of (5.7). To do this, we need to identify a function space $\mathscr{S} \subset L^\infty$

(a) which embeds compactly in L^1_{loc}, in order to be able to show the existence of the limit, at least up to subsequences;
(b) which contains BV, because equations of this type, already in the scalar case, develop singularities along surfaces of codimension one (shocks) in finite time;
(c) such that every transport equation with initial data in $\mathscr{S}$ and with vector field belonging to $\mathscr{S}$ and with bounded divergence possesses a solution belonging to $\mathscr{S}$, to be sure that at each passage of the iteration we can choose a solution which belongs to our space.

The question of the existence of such a space was originally raised by Bressan. However, in a joint work with De Lellis [65], we are able to show that such a space does not exist. We will not be more precise on the statement of this result, for which we refer to [65]. Instead we prefer to describe the idea of our proof and its connection with the counterexample in Section 5.1.

A small modification of the construction by Depauw yields an autonomous divergence-free vector field $u \in L^\infty$, an initial data $\bar{v} \in L^\infty \cap BV$, and a time $T > 0$ such that in $[0, T] \times \mathbb{R}^3$ there exists a unique bounded weak solution of

$$\begin{cases} \partial_t v + \operatorname{div}(uv) = 0 \\ v(0, \cdot) = \bar{v} , \end{cases} \tag{5.8}$$

and $v(t, \cdot)$ converges to $v(T, \cdot)$ weakly* in L^∞ but not strongly in L^1_{loc}.

With an explicit construction it is possible to show the existence of vector fields $f_i \in BV \cap L^\infty$ with $\operatorname{div} f_i \in L^\infty$ and initial data $\bar{w}_i \in BV \cap L^\infty$ such that if we consider the unique weak solutions of

$$\begin{cases} \partial_t w_i + \operatorname{div}(f_i w_i) = 0 \\ w_i(0, \cdot) = \bar{w}_i , \end{cases} \tag{5.9}$$

then $w_i(1, \cdot)$ is the i-th component u_i of Depauw's vector field.

This means that, if we assume the existence of a space $\mathscr{S}$ satisfying our requirements, then necessarily $f_i, \bar{w}_i \in \mathscr{S}$, because of (b). From the closure property (c) and thanks to the fact that the solutions to the systems (5.9) are unique we deduce that $u_i = w_i(1, \cdot) \in \mathscr{S}$. Since, again from (b), $\bar{v} \in \mathscr{S}$, we consider (5.8) and deduce from (c) that $v \in \mathscr{S}$. But we already noticed that $v(t, \cdot)$ converges to $v(T, \cdot)$ weakly* in L^∞ but not strongly in L^1_{loc}, and this violates the compactness condition (a). Hence we deduce that such a space $\mathscr{S}$ cannot exist.

This result is clearly a counterexample to the strategy (it is not possible to solve multi-dimensional systems via reduction to transport equations), but not to the well-posedness result. It would be very interesting to understand whether one can use similar constructions to produce hyperbolic systems of conservation laws $\partial_t U + \text{div} F(U) = 0$ and BV initial data with highly oscillatory admissible solutions. Slight modifications of our example produce fluxes F such that each DF_i is triangular, but the corresponding systems are not hyperbolic.

5.3. Two counterexamples by DiPerna and Lions

In this section we comment two counterexamples to the uniqueness presented by DiPerna and Lions in [84]. These counterexamples are in the spirit of the ODE framework: the "right" notion of solution of the ODE in this relaxed setting will be presented in Section 6.3, but roughly speaking we require that the flow does not concentrate trajectories, in the sense that the push-forward of the Lebesgue measure $\mathscr{L}^d$ via the flow $X(t, \cdot)$ is absolutely continuous with respect to $\mathscr{L}^d$ and has bounded density (this is the so-called regular Lagrangian flow, compare in particular with Definition 6.3.1).

5.3.1. A $W^{1,p}$ vector field with unbounded divergence

The first example of [84] is relative to a vector field with Sobolev regularity but with unbounded divergence, for which an infinite family of solutions to the ODE is constructed. The construction of [84] (taken from a previous work by Beck [30]) is two-dimensional, but since the vector field is constant in the second direction the phenomenon is intrinsically one-dimensional, thus we are going to describe it as a flow on the real line. Moreover, it has been pointed out by Ambrosio that all but one among the constructed flows create concentration of trajectories. Thus this example is not a counterexample to the uniqueness of the regular Lagrangian flow. It would be interesting to understand whether it is possible to have examples of nonuniqueness of the solution even in the cases where the uniqueness of the regular Lagrangian flow is already known.

For any $p \in [1, +\infty[$ fixed, we construct a uniformly continuous vector field $b \in W^{1,p}_{loc} \cap L^\infty(\mathbb{R}; \mathbb{R})$ and infinitely many solutions to the ODE with vector field b. Let us consider a compact set $K \subset [0, 1]$ with empty interior and a function $g \in C^\infty(\mathbb{R})$ with $0 \le g < 1$ and such that $\{g = 0\} = K$. We set $f(x) = \int_0^x g(s)\, ds$. Assuming that $g(s)$ converges to 1 for $|s| \to \infty$ we obtain that f is bijective. We also consider

$$\mathcal{A} = \{\mu \in \mathcal{M}_+(K) \; : \; \mu \text{ does not possess atoms}\}\,.$$

For every $\mu \in \mathcal{A}$ we set

$$\Lambda_\mu^{-1}(x) = x + \mu([0, x])\,, \qquad f_\mu = f \circ \Lambda_\mu\,.$$

Noticing that Λ_μ^{-1} is strictly increasing and continuous we deduce that f_μ is well-defined, continuous and strictly increasing. Moreover, Λ_μ is 1-Lipschitz. We also set

$$b(x) = f'(f^{-1}(x))\,, \qquad X_\mu(t, x) = f_\mu(t + f_\mu^{-1}(x))\,.$$

Since f and f_μ are strictly increasing and continuous, we obtain that b and X_μ are well-defined and continuous. Moreover it is immediate to check that X_μ satisfies the semigroup property.

We want to show that X_μ is everywhere differentiable with respect to t (equivalently, that f_μ is everywhere differentiable) and that

$$\frac{\partial X_\mu}{\partial t}(t, x) = f_\mu'(t + f_\mu^{-1}(x)) = f_\mu'(f_\mu^{-1}(X_\mu(t, x)))$$
$$\stackrel{(*)}{=} f'(f^{-1}(X_\mu(t, x))) = b(X_\mu(t, x))\,. \tag{5.10}$$

Notice that, once we know that f_μ is differentiable, $(*)$ is the only equality in (5.10) which is not an immediate consequence of the definitions given above. The validity of (5.10) will imply that, for every $\mu \in \mathcal{A}$, the map X_μ is a flow relative to the vector field b.

We claim that f_μ is differentiable and that for every $z \in \mathbb{R}$ the following equality holds:

$$f_\mu'(z) = f'(\Lambda_\mu(z))\,. \tag{5.11}$$

Using (5.11) with $z = f_\mu^{-1}(X_\mu(t, x))$ and recalling that $\Lambda_\mu \circ f_\mu^{-1} = f^{-1}$ we see that (5.11) implies the validity of the equality marked by $(*)$ in (5.10). We now prove (5.11) considering two cases.

CASE 1. If $\Lambda_\mu(z) \notin K$, then for s sufficiently close to z we have

$$\Lambda_\mu^{-1}(\Lambda_\mu(z) + s - z) = s\,,$$

since the complement of K is an open set. Hence f_μ is differentiable in z with $f'_\mu(z) = f'(\Lambda_\mu(z))$, and (5.11) holds for such a z.

CASE 2. If $\Lambda_\mu(z) \in K$, recalling that Λ_μ is 1-Lipschitz, for s is sufficiently close to z we obtain

$$|f_\mu(s) - f_\mu(z)| = |f(\Lambda_\mu(s)) - f(\Lambda_\mu(z))| \le C|\Lambda_\mu(s) - \Lambda_\mu(z)|^2 \le C|s-z|^2,$$

since $f'(\Lambda_\mu(z)) = 0$. Hence $f'_\mu(z) = 0 = f'(\Lambda_\mu(z))$ and we eventually get (5.11) also in this case.

We now show how it is possible to choose g in such a way that $b \in W^{1,p}_{\text{loc}}(\mathbb{R}; \mathbb{R})$ for an arbitrary $p \in [1, +\infty[$. Noticing that

$$b'(x) = g'(f^{-1}(x))[g(f^{-1}(x))]^{-1}$$

we can compute

$$\int_\mathbb{R} |b'(x)|^p \, dx = \int_\mathbb{R} |g'(f^{-1}(x))|^p \, |g(f^{-1}(x))|^{-p} \, dx$$

$$= \int_\mathbb{R} |g'(y)|^p \, |g(y)|^{-(p-1)} \, dy.$$

Therefore, it is enough to choose g_0 satisfying all the previous requirements imposed on g and eventually set $g = g_0^p$. Since this implies $|g|^{-(p-1)}|g'|^p = p^p|g_0'|^p$, it is enough to choose g_0 such that $g_0' \in L^p(\mathbb{R})$. Notice also that $\operatorname{div} b(x) = b'(x)$ does not belong to $L^\infty(\mathbb{R})$.

We finally show why the only flow X_μ which does not create concentration of trajectories is the one corresponding to $\mu = 0$. We preliminarily show that

(i) The set $f(K)$ has zero Lebesgue measure;
(ii) The set $\Lambda_\mu^{-1}(K)$ has strictly positive Lebesgue measure, if $\mu \ne 0$.

We observe that the statement in (i) is immediate, since f is smooth and K consists of the critical points of f. To show (ii) we notice that

$$\Lambda_\mu^{-1}([0, 1]) = [0, 1 + m], \tag{5.12}$$

where $m = \mu([0, 1])$. But since the derivative of Λ_μ^{-1} is less or equal to 1 on $[0, 1] \setminus K$ we also have

$$\mathscr{L}^1(\Lambda_\mu^{-1}([0, 1] \setminus K)) \le \mathscr{L}^1([0, 1] \setminus K) \le 1. \tag{5.13}$$

From (5.12) and (5.13) we deduce

$$\mathscr{L}^1(\Lambda_\mu^{-1}(K)) \ge m, \tag{5.14}$$

and this proves (ii).

Using (5.14) and the equivalence

$$X_\mu(t, x) \in f(K) \qquad \Longleftrightarrow \qquad t \in \Lambda_\mu^{-1}(K) - f_\mu^{-1}(x)$$

we eventually obtain the equality

$$\int_\mathbb{R} \int_\mathbb{R} \mathbf{1}_{f(K)}(X_\mu(t, x)) \, dt dx = \int_\mathbb{R} \mathscr{L}^1\left(\Lambda_\mu^{-1}(K) - f_\mu^{-1}(x)\right) dx = +\infty$$

$$(5.15)$$

for every $\mu \in \mathcal{A}$ with $\mu \neq 0$. This implies the existence of two sets A, $B \subset \mathbb{R}$ with strictly positive Lebesgue measure such that

$$X_\mu(t, x) \in f(K) \qquad \text{for every } (t, x) \in A \times B,$$

thus we have concentration of trajectories in the negligible set $f(K)$ for a set of times with positive measure. Moreover, it is possible to check that the density of the measure

$$\left(X_\mu(t, \cdot)_\# \mathscr{L}^1\right) \llcorner \left(\mathbb{R} \setminus f(K)\right)$$

does not belong to $L_{\text{loc}}^\infty(\mathbb{R})$.

This example should be compared with the much simpler one presented in Remark 6.3.3, relative to the square root case. One difference is that in the case presented here each flow X_μ satisfies the semigroup property, while every flow relative to the vector field $b(x) = \sqrt{|x|}$ which "stops" in the origin for a positive time does not satisfy the semigroup property.

5.3.2. A divergence-free vector field with derivatives of fractional order

The second example of [84] regards an autonomous divergence-free vector field b in $\mathbb{R}^2$ such that $b \in W_{\text{loc}}^{s,1}(\mathbb{R}^2; \mathbb{R}^2)$ for every $s \in [0, 1[$ and $b \in L^p(\mathbb{R}^2; \mathbb{R}^2) + L^\infty(\mathbb{R}^2; \mathbb{R}^2)$ for every $p \in [1, 2[$. For the definition and the main properties of Sobolev spaces with fractional order we refer to [1, 124]. We will construct two regular Lagrangian flows (which in fact preserve the Lebesgue measure) relative to this vector field. One might wonder if this lack of derivative which leads to nonuniqueness could be compensated by an enlargement of the summability exponent, *i.e.* if we have uniqueness for vector fields in $W^{s,p}$ with $s < 1$ and $p = p(s)$ sufficiently large. This is not at all the case: some variations of the counterexamples of [5] show that we cannot hope for uniqueness even for vector fields belonging to all the Hölder spaces $C^\alpha(\mathbb{R}^2; \mathbb{R}^2)$ with $\alpha \in [0, 1[$.

Since we want to construct a divergence-free vector field we start by defining the following Hamiltonian function $H : \mathbb{R}^2 \to \mathbb{R}$. We set

$$
H(x, y) = \begin{cases}
-x/|y| & \text{if } |x| < |y| \\
-(x - |y| + 1) & \text{if } x > |y| \\
-(x + |y| - 1) & \text{if } x < -|y| .
\end{cases}
$$

Thus we define b as

$$
\begin{cases}
b_1(x, y) = -\dfrac{\partial H}{\partial y} = -\operatorname{sgn} y \left(\dfrac{x}{|y|^2} \mathbf{1}_{\{|x| \le |y|\}} + \operatorname{sgn} x \; \mathbf{1}_{\{|x| > |y|\}} \right) \\
b_2(x, y) = \dfrac{\partial H}{\partial x} = -\left(\dfrac{1}{|y|} \mathbf{1}_{\{|x| \le |y|\}} + \mathbf{1}_{\{|x| > |y|\}} \right) .
\end{cases}
$$

The only non trivial verification is the $W^{s,1}$ regularity of b: we postpone it to the end of this subsection. We now explicitly define two different flows X and $\tilde{X}$ relative to b. By symmetry it suffices to define the flows for initial data in the set

$$
Q = \{(x, y) \in \mathbb{R}^2 \; : \; x > 0, \quad y > 0, \quad x \ne y \} .
$$

When $x > y$ we set

$$
\begin{cases}
X_1(t, x, y) = \tilde{X}_1(t, x, y) = \begin{cases}
x - t & \text{if } t \le y \\
x - 2y + t & \text{if } t \ge y
\end{cases} \\
X_2(t, x, y) = \tilde{X}_2(t, x, y) = y - t .
\end{cases}
$$

When $x < y$ we set

$$
\begin{cases}
X_1(t, x, y) = \dfrac{x}{y} |y^2 - 2t|^{1/2} \\
X_2(t, x, y) = \sigma |y^2 - 2t|^{1/2}
\end{cases}
$$

and

$$
\begin{cases}
\tilde{X}_1(t, x, y) = \sigma \dfrac{x}{y} |y^2 - 2t|^{1/2} \\
\tilde{X}_2(t, x, y) = \sigma |y^2 - 2t|^{1/2} ,
\end{cases}
$$

where $\sigma = 1$ if $t \le y^2/2$ and $\sigma = -1$ if $t \ge y^2/2$. It is immediate to check that X and $\tilde{X}$ are two different regular Lagrangian flows relative to b and that they preserve the Lebesgue measure in $\mathbb{R}^2$.

Let us finally check that $b \in W^{s,1}_{\text{loc}}(\mathbb{R}^2; \mathbb{R}^2)$. We will give only a sketch of the proof, which uses some technical results from the theory of Besov spaces: see [80, 125] and [124, Section 11], for the precise statements of the results required here.

First of all we observe that

$$B_{1,1}^s(\mathbb{R}^2) \subset W^{s,1}(\mathbb{R}^2),$$

where the Besov space $B_{1,1}^s(\mathbb{R}^2)$ is defined according to [124, Definition 10.3]. Thus it suffices to show that $b \in B_{1,1}^s$ around the origin. The following equivalence of norms holds:

$$\|f\|_{B_{1,1}^s} \sim \|f\|_{B_{1,1}^{s-1}} + \|Df\|_{B_{1,1}^{s-1}}. \tag{5.16}$$

For $s < 1$, the first term in the right hand side of (5.16) can be dropped. Moreover, when $\alpha < 0$, we have the equivalence of norms

$$\|g\|_{B_{1,1}^\alpha} \sim \int_0^1 \epsilon^{-\alpha} \|g * \varphi_\epsilon\|_{L^1} \frac{d\epsilon}{\epsilon}, \tag{5.17}$$

where $\{\varphi_\epsilon\}$ is a standard convolution kernel. Combining all the previous remarks, we need to show that

$$\int_0^1 \epsilon^{1-s} \|Db * \varphi_\epsilon\|_{L^1} \frac{d\epsilon}{\epsilon} < +\infty. \tag{5.18}$$

We estimate the L^1 norm inside the integral as follows:

$$\|Db * \varphi_\epsilon\|_{L^1(\mathbb{R}^2)} = \|Db * \varphi_\epsilon\|_{L^1(B_{2\epsilon}(0))} + \|Db * \varphi_\epsilon\|_{L^1(\mathbb{R}^2 \setminus B_{2\epsilon}(0))}$$
$$\leq \frac{C}{\epsilon} \|b\|_{L^1(B_{3\epsilon}(0))} + C \|Db\|_{\mathcal{M}(\mathbb{R}^2 \setminus B_\epsilon(0))}. \tag{5.19}$$

Inserting (5.19) in (5.18) and noticing that

$$\|b\|_{L^1(B_{3\epsilon}(0))} \sim \epsilon \qquad \text{and} \qquad \|Db\|_{\mathcal{M}(\mathbb{R}^2 \setminus B_\epsilon(0))} \sim \log \epsilon$$

we deduce that the integral in (5.18) is bounded by

$$C \int_0^1 \left[\frac{\log \epsilon}{\epsilon^s} + \frac{1}{\epsilon^s} \right] d\epsilon,$$

which is finite since $s < 1$. This completes the proof.

5.4. Lack of propagation of regularity

In the case $b \in \mathrm{Lip}(\mathbb{R}^d; \mathbb{R}^d)$ we have propagation of the Hölder regularity of the initial data, in the following sense: if $\bar{u} \in C^\alpha \cap L^\infty(\mathbb{R}^d)$ for some

$\alpha \in [0, 1[$ and $u \in L^\infty([0, T] \times \mathbb{R}^d)$ is the unique solution of the Cauchy problem

$$\begin{cases} \partial_t + b \cdot \nabla u = 0 \\ u(0, \cdot) = \bar{u}, \end{cases}$$

then $u(t, \cdot) \in C^\alpha(\mathbb{R}^d)$ for every $t \in [0, T]$. Indeed, the solution can be expressed as

$$u(t, x) = \bar{u}(X(t, \cdot)^{-1}(x))$$

and the inverse of the flow $X(t, \cdot)^{-1} : \mathbb{R}^d \to \mathbb{R}^d$ is a Lipschitz map (Corollary 1.3.2).

One might wonder if similar results are true for less regular vector fields. In fact, for "classical" regularities of the solution, nothing of this sort is true, as pointed out by Colombini, Luo and Rauch [62]. However, in Section 7.5 we will show that a mild regularity, namely Lipexp$_p$ regularity, is propagated by $W^{1,p}$ vector fields with bounded divergence. In the remaining of this section we illustrate the two main examples of [62].

Example 5.4.1 (Lack of propagation of continuity). Consider a non-negative function $f \in C(\mathbb{R})$ such that $f(s) = 0$ for $s \leq 0$ and $f(s) = s(\log s)^2$ for $0 < s \leq e^{-2}/2$. Assume that f is nondecreasing and uniformly bounded. Set

$$b(x, y) = (-f(y), -f(x)).$$

Then b is a bounded divergence-free vector field, which belongs to $C^\alpha(\mathbb{R}^2; \mathbb{R}^2)$ for every $\alpha \in [0, 1[$ and to $W^{1,p}_{loc}(\mathbb{R}^2; \mathbb{R}^2)$ for every $p \in [1, +\infty[$. Now consider any initial data $\bar{u} \in C_c^\infty(\mathbb{R}^2)$ such that $\bar{u} = 0$ if $x \leq 0$ and $y \leq 0$, and $\bar{u} > 0$ if $x > 0$ and $y > 0$ are small. Then, for every relatively open set $\Omega \subset [0, T] \times \mathbb{R}^2$ such that $(0, 0, 0) \in \Omega$, the unique solution $u \in L^\infty([0, T] \times \mathbb{R}^2)$ with initial data $\bar{u}$ is not continuous in Ω.

Example 5.4.2 (Lack of propagation of BV regularity). Consider a function $g \in C_c(\mathbb{R})$ such that $g(s) = -s \log |s| + s$ for s close to 0 and set

$$b(x, y) = (g(y), 0).$$

Set $r = \sqrt{x^2 + y^2}$ and consider an initial data $\bar{u} \in L^\infty \cap W^{1,1}(\mathbb{R}^2)$ such that

$$\bar{u}(x, y) = \cos\left(\frac{1}{r(\log r)^2}\right) \qquad \text{for } r \text{ small.}$$

Then there exists a unique solution $u \in L^\infty([0, T] \times \mathbb{R}^2)$ with initial data $\bar{u}$, and $u(t, \cdot) \notin BV_{loc}(\mathbb{R}^2)$ for t small.

Part II

The Lagrangian
viewpoint

Chapter 6
The connection between PDE and ODE

In this chapter we are going to describe the connection between the Eulerian problem (PDE) and the Lagrangian problem (ODE) out of the smooth setting. In Section 1.4 we presented the classical theory of characteristics, which relates the two problems in the case of a sufficiently smooth vector field. Here we are going to present the theory of *regular Lagrangian flows*, developed by Ambrosio in [8] (see also [13]). In order to study existence, uniqueness and stability of solutions to the ODE, we consider suitable measures in the space of continuous maps, allowing for superposition of trajectories. Then, in some special situations we are able to show that this superposition actually does not occur, but still this "probabilistic" interpretation is very useful to understand the underlying techniques and to give an intrinsic characterization of the flow. We will establish an abstract connection between the well-posedness in the class of bounded solutions for the continuity equation

$$\begin{cases} \partial_t \mu + \operatorname{div}(b\mu) = 0 \\ \mu_0 = \bar{\mu} \end{cases} \tag{6.1}$$

and the well-posedness of a suitable notion of solution (the regular Lagrangian flow) of the ordinary differential equation

$$\begin{cases} \dot{\gamma}(t) = b(t, \gamma(t)) \\ \gamma(0) = x . \end{cases} \tag{6.2}$$

With "abstract" we mean that the spirit of these results will be to deduce the well-posedness for the ODE from the well-posedness for the PDE; hence all the results shown in the previous chapters will imply existence, uniqueness and stability for the regular Lagrangian flow.

In Section 6.5 we briefly comment on the notion of flow defined by DiPerna and Lions. Section 6.6 is devoted to the case of nearly incompressible vector fields: we follow an approach, due to De Lellis, based on the notion of density of the regular Lagrangian flow. The last section presents Bressan's compactness conjecture and some recent advances relative to it.

6.1. Pointwise uniqueness and measure valued solutions

We first present a very general criterion, which relates the pointwise uniqueness for the ordinary differential equation with the uniqueness for positive measure-valued solutions to the continuity equation. Notice that, in order to give a meaning to the product $b\mu$ when μ is a measure, we assume b to be defined *everywhere* in $[0, T] \times \mathbb{R}^d$.

Theorem 6.1.1. *Let $A \subset \mathbb{R}^d$ be a Borel set. Then the following two properties are equivalent:*

(i) *Solutions of the ordinary differential equation (6.2) are unique for every initial point $x \in A$;*
(ii) *Positive measure-valued solutions of the continuity equation (6.1) are unique for every initial data $\bar{\mu}$ which is a positive measure concentrated on A, i.e. such that $\bar{\mu}(\mathbb{R}^d \setminus A) = 0$.*

Proof of (ii) $\Rightarrow$ (i) *in Theorem* 6.1.1. This implication is rather easy. Assume that for some $x \in A$ there exist two different solutions $\gamma(t)$ and $\tilde{\gamma}(t)$ of the ODE starting from x. Then consider $\mu_t = \delta_{\gamma(t)}$ and $\tilde{\mu}_t = \delta_{\tilde{\gamma}(t)}$. We clearly have $\mu_0 = \tilde{\mu}_0 = \delta_x$. It is readily checked that μ_t and $\tilde{\mu}_t$ are solutions of the continuity equation, but since they are different we are violating assumption (ii). $\qquad\square$

The converse implication is much more complicated and requires the so-called *superposition principle*, which roughly speaking says that every positive measure-valued solution of the continuity equation can be obtained as a superposition of solutions obtained via propagation along characteristics. The superposition principle will be presented in the next section, together with the proof of the implication (i) $\Rightarrow$ (ii) of Theorem 6.1.1.

We remark that the applicability of Theorem 6.1.1 is in fact very limited. On one hand, pointwise uniqueness for the ordinary differential equation is known only under very strong regularity assumptions on the vector field, namely in the cases we presented in Chapter 1 (for instance Lipschitz regularity, one-sided Lipschitz condition, Osgood condition). On the other hand, uniqueness for the continuity equation is known only for particular classes of solutions, tipically for solutions which are bounded functions. It is reasonable that this kind of "weaker PDE uniqueness" should reflect into a weaker notion of uniqueness for the ODE: this leads to the concept of *regular Lagrangian flow*, which is presented in Section 6.3.

6.2. The superposition principle

In order to understand better the meaning of the superposition principle we recall formula (1.17). If there is a unique flow $X(t, x)$ associated to the vector field b, the only solution of the continuity equation with initial data $\bar{\mu} \in \mathcal{M}(\mathbb{R}^d)$ is the measure μ_t characterized by

$$\langle \mu_t, \varphi \rangle = \int_{\mathbb{R}^d} \varphi(X(t, x)) \, d\bar{\mu}(x) \qquad \forall \varphi \in C_c(\mathbb{R}^d) . \qquad (6.3)$$

In the following we use the notation Γ_T for the space $C([0, T]; \mathbb{R}^d)$ of continuous paths in $\mathbb{R}^d$. For every $x \in \mathbb{R}^d$ let us consider a probability measure $\eta_x \in \mathcal{P}(\Gamma_T)$ concentrated on the trajectories $\gamma \in \Gamma_T$ which are absolutely continuous integral solutions of the ordinary differential equation with $\gamma(0) = x$. All the families $\{\eta_x\}_{x \in \mathbb{R}^d}$ in the following discussions are weakly measurable, *i.e.* for every function $\Phi \in C_b(\Gamma_T)$ the map

$$x \mapsto \langle \eta_x, \Phi \rangle = \int_{\Gamma_T} \Phi(\gamma) \, d\eta_x(\gamma)$$

is measurable.

Definition 6.2.1 (Superposition solution). The *superposition solution* induced by the family $\{\eta_x\}_{x \in \mathbb{R}^d}$ is the family of measures $\mu_t^{\eta_x} \in \mathcal{M}(\mathbb{R}^d)$, for $t \in [0, T]$, defined as follows:

$$\langle \mu_t^{\eta_x}, \varphi \rangle = \int_{\mathbb{R}^d} \left(\int_{\Gamma_T} \varphi(\gamma(t)) \, d\eta_x(\gamma) \right) d\bar{\mu}(x) \qquad \forall \varphi \in C_c(\mathbb{R}^d) . \qquad (6.4)$$

Using this notation we can give an alternative interpretation of (6.3). If for every $x \in \mathbb{R}^d$ the solution of the ODE starting from x is unique, then the only admissible measure η_x in (6.4) is $\eta_x = \delta_{X(\cdot, x)}$. But then we have

$$\langle \mu_t^{\delta_{X(\cdot, x)}}, \varphi \rangle = \int_{\mathbb{R}^d} \left(\int_{\Gamma_T} \varphi(\gamma(t)) \, d\delta_{X(\cdot, x)}(\gamma) \right) d\bar{\mu}(x)$$

$$= \int_{\mathbb{R}^d} \varphi(X(t, x)) \, d\bar{\mu}(x) ,$$

so in this case we reduce to the "deterministic" formula (6.3). We can regard the superposition solution of Definition 6.2.1 as a "probabilistic" version of (6.3): if there is more than one solution to the ordinary differential equation, then we define our "averaged push-forward" by substituting the quantity $\varphi(X(t,x))$ with the average $\int_{\Gamma_T}\varphi(\gamma(t))\,d\eta_x(\gamma)$. Let us check that (6.4) defines a solution of the continuity equation. Since the measure η_x is concentrated on solutions of the ODE starting from $x\in\mathbb{R}^d$ for $\mathscr{L}^d$-a.e. $x\in\mathbb{R}^d$, using Fubini's theorem we deduce that

$$\dot\gamma(t)=b(t,\gamma(t))\qquad\text{for }\mathscr{L}^d\otimes\eta_x\text{-a.e. }(x,\gamma)\in\mathbb{R}^d\times\Gamma_T$$

for $\mathscr{L}^1$-a.e. $t\in[0,T]$. We notice that, for any $\varphi\in C_c^\infty(\mathbb{R}^d)$, the map $t\mapsto\langle\mu_t^{\eta_x},\varphi\rangle$ is Lipschitz. Indeed, we can estimate

$$\left|\varphi(\gamma(t))-\varphi(\gamma(s))\right|\le\|\nabla\varphi\|_\infty\|b\|_\infty|t-s|$$

and this implies

$$\left|\langle\mu_t^{\eta_x},\varphi\rangle-\langle\mu_s^{\eta_x},\varphi\rangle\right|=\left|\int_{\mathbb{R}^d}\int_{\Gamma_T}\left[\varphi(\gamma(t))-\varphi(\gamma(s))\right]d\eta_x(\gamma)\,d\bar\mu(x)\right|$$

$$\le\|\nabla\varphi\|_\infty\|b\|_\infty\bar\mu\big(\operatorname{spt}\varphi+B_{T\|b\|_\infty}(0)\big)|t-s|.$$

Thus the distributional derivative of the map $t\mapsto\langle\mu_t^{\eta_x},\varphi\rangle$ coincides with the pointwise one. But since an immediate computation of the pointwise derivative gives

$$\frac{d}{dt}\langle\mu_t^{\eta_x},\varphi\rangle=\frac{d}{dt}\int_{\mathbb{R}^d}\left(\int_{\Gamma_T}\varphi(\gamma(t))\,d\eta_x(\gamma)\right)d\bar\mu(x)$$

$$=\int_{\mathbb{R}^d}\int_{\Gamma_T}\nabla\varphi(\gamma(t))\cdot b(t,\gamma(t))\,d\eta_x(\gamma)\,dx$$

$$=\int_{\mathbb{R}^d}b(t,x)\cdot\nabla\varphi(x)\,d\mu_t(x),$$

we obtain that the continuity equation holds.

The superposition principle says that, for positive solutions, this construction can be reversed: every positive measure-valued solution μ_t can be realized as a superposition solution $\mu_t^{\eta_x}$ for some $\{\eta_x\}_{x\in\mathbb{R}^d}$ as above. The result in the form we are going to present is a slight generalization of the one proved in [18, Section 8.2]: since we will need to deal with L^∞ solutions of the continuity equation we need to consider here measures μ_t which are only locally finite. We refer to [13, Section 4] for the case in which $\mu_t \in \mathcal{P}(\mathbb{R}^d)$ and the vector field satisfies the global summability assumption

$$\int_0^T \int_{\mathbb{R}^d} \frac{|b(t,x)|}{1+|x|}\, d\mu_t(x)\, dt < +\infty.$$

See also Maniglia [111] for an extension to the non homogeneous case.

Theorem 6.2.2 (Superposition principle). *Fix a bounded vector field* $b : [0, T] \times \mathbb{R}^d \to \mathbb{R}^d$ *and let* $\mu_t \in \mathcal{M}_+(\mathbb{R}^d)$ *be a positive locally finite measure-valued solution of the continuity equation. Then* μ_t *is a superposition solution, i.e. there exists a family* $\{\eta_x\}_{x \in \mathbb{R}^d} \subset \mathcal{P}(\Gamma_T)$, *with* η_x *concentrated on absolutely continuous integral solutions of the ODE starting from* x, *for* $\bar{\mu}$-*a.e.* $x \in \mathbb{R}^d$, *such that* $\mu_t = \mu_t^{\eta_x}$ *for any* $t \in [0, T]$.

Proof. STEP 1.   UNIFORM CONTROL OF THE LOCAL MASSES.   The measure μ_t is just locally finite; however, due to the finite speed of propagation of the transport equation (we are assuming that b is uniformly bounded), it is easy to show the existence of a function m_R, independent of the time, such that

$$\mu_t(B_R(0)) \leq m_R \qquad \text{for every } t \in [0, T]. \qquad (6.5)$$

This can be proved by integration on suitable cones, see for instance [11, Lemma 2.11] (the argument used for bounded solutions remains unchanged in the case of measure-valued solutions).

STEP 2.   CONSTRUCTION OF AN ADAPTED CONVOLUTION KERNEL. We want to contruct a positive convolution kernel $\rho \in C^k(\mathbb{R}^d)$ with spt $\rho = \mathbb{R}^d$ and $\int_{\mathbb{R}^d} \rho(x)\, dx = 1$ in such a way that, for some function $\tilde{m}_R$, we have

$$\mu_t^\epsilon(B_R(0)) \leq \tilde{m}_R \qquad \text{for every } t \in [0, T] \text{ and every } \epsilon \in]0, 1[, \quad (6.6)$$

where $\mu_t^\epsilon = \mu_t * \rho^\epsilon$. This can be done as follows.

For every $j \in \mathbb{N}$ consider a standard convolution kernel ρ_j with spt $\rho_j = B_j(0)$, $\rho_j \geq 0$ and $\int_{\mathbb{R}^d} \rho_j(x)\, dx = 1$. We define

$$\rho(x) = \sum_{j=1}^\infty c_j \rho_j(x).$$

We want to choose the numbers c_j in order to obtain a convolution kernel ρ which satisfies the desired properties. If we choose $c_j > 0$ for every j and $\sum_j c_j = 1$ we obtain that ρ is strictly positive on the whole $\mathbb{R}^d$ and has unit total mass. Moreover, in order to obtain a C^k function, we also require $\sum_j c_j \|\rho_j\|_{C^k} < \infty$, which is clearly satisfied for every $k \in \mathbb{N}$ if we choose a sequence c_j which goes to zero sufficiently fast.

We now show how it is possible to choose the sequence c_j in such a way that the uniform control (6.6) is satisfied. We compute

$$\mu_t^\epsilon(B_R(0)) = \int_{B_R(0)} \mu_t^\epsilon(x)\,dx = \int_{B_R(0)} \int_{\mathbb{R}^d} \rho^\epsilon(x-y)\,d\mu_t(y)\,dx$$

$$= \int_{B_R(0)} \int_{\mathbb{R}^d} \sum_{j=1}^\infty c_j \rho_j^\epsilon(x-y)\,d\mu_t(y)\,dx$$

$$= \sum_{j=1}^\infty c_j \int_{B_R(0)} \int_{\mathbb{R}^d} \rho_j^\epsilon(x-y)\,d\mu_t(y)\,dx$$

$$= \sum_{j=1}^\infty c_j \int_{\mathbb{R}^d} \int_{B_R(0)} \rho_j^\epsilon(x-y)\,dx\,d\mu_t(y)$$

$$= \sum_{j=1}^\infty c_j \int_{B_{R+j}(0)} \int_{B_R(0)} \rho_j^\epsilon(x-y)\,dx\,d\mu_t(y)$$

$$\leq \sum_{j=1}^\infty c_j \int_{B_{R+j}(0)} d\mu_t(y) \leq \sum_{j=1}^\infty c_j m_{R+j}.$$

This means that we need to choose the sequence $c_j > 0$ in such a way that

$$\tilde{m}_R = \sum_{j=1}^\infty c_j m_{R+j} < \infty \tag{6.7}$$

for every $R \in \mathbb{N}$. It can be easily checked that it is enough to choose each c_j such that

$$c_j m_j, \ c_j m_{j+1}, \ \dots \ c_j m_{2j} \leq \frac{1}{2^j}.$$

Indeed with this choice we can estimate

$$\tilde{m}_R = \sum_{j=1}^\infty c_j m_{R+j} = \sum_{j=1}^{R-1} c_j m_{R+j} + \sum_{j=R}^\infty c_j m_{R+j}$$

$$\leq \sum_{j=1}^{R-1} c_j m_{R+j} + \sum_{j=R}^\infty \frac{1}{2^j} < \infty.$$

Noticing that, if a sequence c_j satisfies (6.7) and $0 < c'_j \le c_j$, then also the sequence c'_j satisfies (6.7), we conclude that it is possible to choose a sequence which satisfies all the requirements we need. Hence we have constructed a kernel ρ with the claimed property (6.6).

STEP 3. SMOOTHING. We fix a convolution kernel ρ as in the previous step and we consider

$$\mu_t^\epsilon(x) = (\mu_t * \rho^\epsilon)(x) \qquad \text{and} \qquad b^\epsilon(t, x) = \frac{(b(t, \cdot)\mu_t) * \rho^\epsilon(x)}{\mu_t^\epsilon(x)}.$$

It is immediate to check the uniform L^∞ bound

$$|b^\epsilon(t, x)| = \frac{|(b(t, \cdot)\mu_t) * \rho^\epsilon(x)|}{\mu_t^\epsilon(x)}$$

$$\le \frac{\int_{\mathbb{R}^d} |\rho^\epsilon(x - y)b(t, y)| \, d\mu_t(y)}{\mu_t^\epsilon(x)} \qquad (6.8)$$

$$\le \frac{\|b\|_\infty \mu_t^\epsilon(x)}{\mu_t^\epsilon(x)} = \|b\|_\infty$$

and that μ_t^ϵ is a solution of the continuity equation with vector field b^ϵ, that is

$$\partial_t \mu^\epsilon + \operatorname{div}(b^\epsilon \mu_t^\epsilon) = \partial_t \mu_t * \rho^\epsilon + \operatorname{div}(b(t, \cdot)\mu_t) * \rho^\epsilon = 0.$$

Since b^ϵ is globally bounded and C^k with respect to the space (it is indeed a ratio of two C^k functions and the denominator does not vanish) we can define a unique flow $X^\epsilon(t, x)$ associated to b^ϵ, which is globally defined for $t \in [0, T]$. The representation $\mu_t^\epsilon = X^\epsilon(t, \cdot)_\# \mu_0^\epsilon$ given by Proposition 1.4.3 holds and we define

$$\eta_x^\epsilon = \delta_{X^\epsilon(\cdot, x)} \in \mathcal{P}(\Gamma_T)$$

and

$$\eta^\epsilon = \mu_0^\epsilon \otimes \eta_x^\epsilon \in \mathcal{M}_+(\mathbb{R}^d \times \Gamma_T),$$

where the tensor product by definition satisfies

$$\langle \eta^\epsilon, \Phi(x, \gamma) \rangle = \int_{\mathbb{R}^d} \int_{\Gamma_T} \Phi(x, \gamma) \, d\eta_x^\epsilon(\gamma) \, d\mu_0^\epsilon(x)$$

for every function $\Phi \in C_b(\mathbb{R}^d \times \Gamma_T)$ whose support has compact projection over $\mathbb{R}^d$.

STEP 4. TIGHTNESS. We consider the functional $\Psi : \mathbb{R}^d \times \Gamma_T \rightarrow [0, +\infty]$ defined by

$$\Psi : (x, \gamma) \mapsto \int_0^T |\dot{\gamma}(t)|^2 \, dt \, ;$$

the functional takes the value $+\infty$ for all the curves $\gamma \in \Gamma_T$ which do not belong to $AC^2([0, T]; \mathbb{R}^d)$, the space of absolutely continuous maps with square-integrable derivative. For every $R > 0$ we define the space $\Gamma_{T,R} \subset \Gamma_T$ of the curves starting from $B_R(0)$, that is

$$\Gamma_{T,R} = \{\gamma \in \Gamma_T : \gamma(0) \in B_R(0)\},$$

and the localized functional

$$\Psi^R(x, \gamma) = \begin{cases} \Psi(x, \gamma) & \text{if } (x, \gamma) \in B_R(0) \times \Gamma_{T,R} \\ +\infty & \text{otherwise.} \end{cases}$$

It is readily checked that Ψ^R is coercive for every R. For every finite $\lambda > 0$ let $(x, \gamma) \in \{\Psi^R \leq \lambda\}$. Since λ is finite we deduce that $x \in B_R(0)$ and that $\gamma(0) \in B_R(0)$. Moreover for every s and $t \in [0, T]$ we can compute

$$|\gamma(t) - \gamma(s)| = \left| \int_s^t \dot{\gamma}(\tau) \, d\tau \right|$$

$$\leq \left(\int_0^T |\dot{\gamma}(\tau)|^2 \, d\tau \right)^{1/2} |t - s|^{1/2} \leq \lambda^{1/2} |t - s|^{1/2}.$$

Applying the Ascoli–Arzelà theorem we deduce that each sublevel is relatively compact in $\mathbb{R}^d \times \Gamma_T$, as desired.

We define also the truncated measure

$$\eta^{\epsilon,R} = \eta^\epsilon \llcorner \left(B_R(0) \times \Gamma_{T,R} \right) \in \mathcal{M}_+(\mathbb{R}^d \times \Gamma_T),$$

which is in fact a finite measure. Now we want to evaluate the value of the truncated measure $\eta^{\epsilon,R}$ on the localized functional Ψ^R. Using the definition of η^ϵ, the fact that μ_t^ϵ is a solution of the continuity equation and the uniform bound on b^ϵ we can estimate

$$\langle \eta^{\epsilon,R}, \Psi^R \rangle = \int_{\mathbb{R}^d \times \Gamma_T} \Psi^R(x, \gamma) \, d\eta^{\epsilon,R}(x, \gamma)$$

$$= \int_{B_R(0)} \int_{\Gamma_{T,R}} \Psi(x, \gamma) \, d\eta_x^\epsilon(\gamma) \, d\mu_0^\epsilon(x)$$

$$= \int_{B_R(0)} \int_0^T \left| \frac{\partial X^\epsilon}{\partial t}(t, x) \right|^2 dt \, d\mu_0^\epsilon(x)$$

$$= \int_0^T \int_{B_R(0)} |b^\epsilon(t, X^\epsilon(t, x))|^2 \, d\mu_0^\epsilon(x) \, dt$$

$$= \int_0^T \int_{X^\epsilon(t, B_R(0))} |b^\epsilon(t, x)|^2 \, d\mu_t^\epsilon(x) \, dt$$

$$\leq \int_0^T \int_{B_{R+T\|b\|_\infty}(0)} |b^\epsilon(t, x)|^2 \, d\mu_t^\epsilon(x) \, dt$$

$$\leq \|b\|_{L^\infty([0,T] \times \mathbb{R}^d)}^2 \int_0^T d\mu_t^\epsilon \left(B_{R+T\|b\|_\infty}(0) \right) dt$$

$$\leq T \|b\|_{L^\infty([0,T] \times \mathbb{R}^d)}^2 \tilde{m}_{R+T\|b\|_\infty} .$$

Together with the coercivity of Ψ^R, this gives that, for every fixed R, the family $\{\eta^{\epsilon,R}\}_\epsilon \subset \mathcal{M}_+(\mathbb{R}^d \times \Gamma_T)$ is tight (see the discussion in Appendix A.1). Noticing that this family is also equi-bounded by the simple estimate

$$\eta^{\epsilon,R}(\mathbb{R}^d \times \Gamma_T) = \mu_0^\epsilon(B_R(0)) \leq \tilde{m}_R ,$$

we can apply the Prokhorov theorem (Theorem A.1.1) and deduce that it is relatively sequentially narrowly compact. With a standard diagonal argument (based on the fact that narrow convergence is preserved under the restriction of measures) it is possible to construct a sequence $\{\eta^{\epsilon_i}\}_i \subset \mathcal{M}_+(\mathbb{R}^d \times \Gamma_T)$ and a measure $\eta \in \mathcal{M}_+(\mathbb{R}^d \times \Gamma_T)$ such that

$$\eta^{\epsilon_i,R} \to \eta^R \qquad \text{narrowly in } \mathcal{M}_+(\mathbb{R}^d \times \Gamma_T) \text{ for every } R.$$

For future use, we say that η^{ϵ_i} converges toward η in the sense of the truncated narrow convergence in $\mathcal{M}_+(\mathbb{R}^d \times \Gamma_T)$.

STEP 5. DISINTEGRATION AND EQUALITY $\mu_t = \mu_t^{\eta_x}$. Up to now we have constructed a measure $\eta \in \mathcal{M}_+(\mathbb{R}^d \times \Gamma_T)$. It is clear that the marginal on $\mathbb{R}^d$ of η is $\bar\mu$. Indeed, for every $\epsilon \in]0, 1[$ we have $(\pi_{\mathbb{R}^d})_\#\eta^\epsilon = \mu_0^\epsilon$ and the truncated narrow convergence is inherited by the marginal; this, combined with the fact that μ_0^ϵ converges to $\bar\mu$ weakly in $\mathbb{R}^d$, gives that $(\pi_{\mathbb{R}^d})_\#\eta = \bar\mu$. This implies that we can apply the disintegration theorem recalled in Appendix A.1 to obtain

$$\eta = \bar\mu \otimes \eta_x ,$$

with $\eta_x \in \mathcal{P}(\Gamma_T)$ for $\bar\mu$-a.e. $x \in \mathbb{R}^d$.

We notice that in the proof we preferred to show convergence at the level of measures on the product $\mathbb{R}^d \times \Gamma_T$ rather than convergence of η_x for $\bar\mu$-a.e. x. In this way we get a subsequence $\{\epsilon_i\}$ along which we have convergence which does not depend on the point x; choosing the limit η_x along a different subsequence for each point x could produce a family $\{\eta_x\}_{x\in\mathbb{R}^d}$ with non-measurable dependence with respect to x.

We now check that $\mu_t^{\eta_x} = \mu_t$. Indeed, for every $\epsilon \in]0, 1[$, applying the definition of η_x^ϵ and the fact that $\mu_t^\epsilon = X^\epsilon(t, \cdot)_\#\mu_0^\epsilon$, we deduce that

$$\langle \mu_t^{\eta_x^\epsilon}, \varphi \rangle = \int_{\mathbb{R}^d} \varphi(X^\epsilon(t, x)) \, d\mu_0^\epsilon(x) = \int_{\mathbb{R}^d} \varphi(x) \, d\mu_t^\epsilon(x) = \langle \mu_t^\epsilon, \varphi \rangle \quad (6.9)$$

for every $\varphi \in C_c(\mathbb{R}^d)$. Note that

$$\int_{\mathbb{R}^d} \varphi(x) d\mu_t^\epsilon(x) \to \int_{\mathbb{R}^d} \varphi(x) d\mu_t(x) = \langle \mu_t, \varphi \rangle \quad (6.10)$$

by the truncated narrow convergence of μ_t^ϵ to μ_t. Moreover

$$\int_{\mathbb{R}^d} \varphi(X^\epsilon(t, x)) \, d\mu_0^\epsilon(x) = \int_{\mathbb{R}^d \times \Gamma_T} \varphi(\gamma(t)) \, d\eta^\epsilon(x, \gamma)$$
$$\to \int_{\mathbb{R}^d \times \Gamma_T} \varphi(\gamma(t)) d\eta(x, \gamma) = \int_{\mathbb{R}^d} \int_{\Gamma_T} \varphi(\gamma(t)) d\eta_x(\gamma) d\bar\mu(x) = \langle \mu_t^{\eta_x}, \varphi \rangle$$
$$(6.11)$$

by the truncated narrow convergence of η^ϵ to η along the chosen subsequence. We finally obtain the desired equality $\mu_t^{\eta_x} = \mu_t$ from (6.10) and (6.11).

STEP 6. η IS CONCENTRATED ON SOLUTIONS OF THE ODE. In this last step we want to show that η_x is concentrated on solutions of the ODE for $\bar\mu$-a.e. $x \in \mathbb{R}^d$. It is enough to show that

$$\int_{B_R(0)\times\Gamma_T} \left| \gamma(t) - x - \int_0^t b(s, \gamma(s)) \, ds \right| d\eta(x, \gamma) = 0 \quad (6.12)$$

for every $t \in [0, T]$ and for every $R > 0$. Indeed, for every $t \in [0, T]$ and every $R \in \mathbb{N}$ this gives a $\bar{\mu}$-negligible set $N_{t,R} \subset B_R(0)$ such that the equality

$$\gamma(t) = x + \int_0^t b(s, \gamma(s))\, ds \tag{6.13}$$

holds for every $x \in B_R(0) \setminus N_{t,R}$ for η_x-a.e. $\gamma \in \Gamma_T$. Choosing

$$N = \bigcup_{t \in [0,T] \cap \mathbb{Q}} \bigcup_{R \in \mathbb{N}} N_{t,R}$$

(which is clearly $\mathscr{L}^d$-negligible) and using the continuity of each trajectory γ, we deduce that (6.13) holds for every $t \in [0, T]$, for every $x \in \mathbb{R}^d \setminus N$ for η_x-a.e. $\gamma \in \Gamma_T$.

Our objective is then to show (6.12) for any $t \in [0, T]$ and any $R \in \mathbb{N}$. The technical difficulty is that this test function, due to the lack of regularity of b, is not continuous. We claim that

$$\int_{B_R(0) \times \Gamma_T} \left| \gamma(t) - x - \int_0^t a(s, \gamma(s))\, ds \right| d\eta(x, \gamma)$$

$$\tag{6.14}$$

$$\leq \int_0^T \int_{\mathbb{R}^d} |b(s, x) - a(s, x)|\, d\mu_s(x)\, ds$$

for every continuous vector field a. We first show how the claimed equation (6.14) implies our thesis. It is enough to choose a sequence $\{a_k\}$ of continuous vector fields such that

$$\int_0^T \int_{\mathbb{R}^d} |b(s, x) - a_k(s, x)|\, d\mu_s(x)\, ds \to 0 \qquad \text{as } k \to \infty. \tag{6.15}$$

This can be done by observing that continuous functions are dense in $L^1_{\text{loc}}([0, T] \times \mathbb{R}^d), \mu)$ with respect to the global norm $L^1([0, T] \times \mathbb{R}^d, \mu)$ (here $\mu = \int_0^T \mu_t\, dt \in \mathcal{M}_+([0, T] \times \mathbb{R}^d)$). To see this, it is enough to write $f \in L^1_{\text{loc}}([0, T] \times \mathbb{R}^d, \mu)$ as a sum of compactly supported functions f_i and then mollify each f_i with a fixed kernel β^ϵ, but choosing a parameter $\epsilon = \epsilon(i)$ small enough in such a way that the difference $f_i - f_i * \beta^{\epsilon(i)}$ decays like 2^{-i} and the series of the characteristic functions of spt $\left(f_i * \beta^{\epsilon(i)}\right)$ is bounded.

Assuming the validity of (6.14) we can then estimate

$$\int_{B_R(0)\times\Gamma_T} \left| \gamma(t) - x - \int_0^t b(s,\gamma(s))\,ds \right| d\eta(x,\gamma)$$

$$\leq \int_{B_R(0)\times\Gamma_T} \left| \gamma(t) - x - \int_0^t a_k(s,\gamma(s))\,ds \right|$$

$$+ \left| \int_0^t b(s,\gamma(s)) - a_k(s,\gamma(s))\,ds \right| d\eta(x,\gamma)$$

$$\leq 2\int_0^T \int_{\mathbb{R}^d} |b(s,x) - a_k(s,x)|\,d\mu_s(x)\,ds.$$

In the last inequality we used (6.14) to estimate the first term and the fact that $\mu_t = \mu_t^{\eta x}$ to compute the second one (which is eventually estimated by the global integral on the whole $\mathbb{R}^d$). Recalling (6.15) we obtain the desired formula (6.12).

It remains to show (6.14). For every $\epsilon \in]0,1[$ we compute

$$\int_{B_R(0)\times\Gamma_T} \left| \gamma(t) - x - \int_0^t a(s,\gamma(s))\,ds \right| d\eta^\epsilon(x,\gamma)$$

$$= \int_{B_R(0)} \left| X^\epsilon(t,x) - x - \int_0^t a(s,X^\epsilon(s,x))\,ds \right| d\mu_0^\epsilon(x)$$

$$\leq \int_0^t \int_{B_R(0)} |(b^\epsilon - a)(s,X^\epsilon(s,x))|\,d\mu_0^\epsilon(x)\,ds$$

$$\leq \int_0^t \int_{B_R(0)} \left[|(b^\epsilon-a^\epsilon)(s,X^\epsilon(s,x))| + |(a^\epsilon-a)(s,X^\epsilon(s,x))| \right] d\mu_0^\epsilon(x)\,ds$$

$$\leq \int_0^t \int_{B_{R+T\|b\|_\infty}(0)} \left[|(b^\epsilon - a^\epsilon)(s,x)| + |(a^\epsilon - a)(s,x)| \right] d\mu_s^\epsilon(x)\,ds,$$

where we have used the definition of η^ϵ, the fact that $\mu_t^\epsilon = X^\epsilon(t,\cdot)_\# \mu_0^\epsilon$ and we have set $a^\epsilon(t,x) = [(a(t,\cdot)\mu_t) * \rho^\epsilon(x)]/\mu_t^\epsilon(x)$. We estimate the first integral. Setting $d = b - a$ we notice that

$$|b^\epsilon - a^\epsilon| = |d^\epsilon| = \frac{|(d(t,\cdot)\mu_t) * \rho^\epsilon|}{\mu_t^\epsilon},$$

hence

$$\int_0^t \int_{B_{R+T\|b\|_\infty}(0)} |d^\epsilon(s,x)| \, d\mu_s^\epsilon(x) \, ds$$

$$= \int_0^t \int_{B_{R+T\|b\|_\infty}(0)} \left| \left(d(s,\cdot)\mu_s \right) * \rho^\epsilon(x) \right| \, dx \, ds$$

$$\leq \int_0^t \int_{\mathbb{R}^d} \int_{B_{R+T\|b\|_\infty}(0)} \rho^\epsilon(x-y) \, dx |d(s,y)| \, d\mu_s(y) \, ds$$

$$\leq \int_0^T \int_{\mathbb{R}^d} |d(s,y)| \, d\mu_s(y) \, ds .$$

Since a is continuous we have that $a^\epsilon \to a$ locally uniformly, hence the second integral vanishes as $\epsilon \to 0$. Passing to the limit in the inequalities above (along the subsequence for which we have truncated narrow converge of η^ϵ to η) we obtain (6.14). □

Using the superposition principle we can conclude the proof of Theorem 6.1.1.

Proof of (i)⇒(ii) *in Theorem* 6.1.1. Let μ_t be a positive measure-valued solution of the continuity equation with initial data $\bar{\mu}$. Applying Theorem 6.2.2 we deduce that $\mu_t = \mu_t^{\eta_x}$, with $\eta_x \in \mathcal{P}(\Gamma_T)$ concentrated on the absolutely continuous integral solutions of the ODE starting from x, for every point $x \in A$. But assumption (i) precisely means that, for every $x \in A$, the solution is unique. Hence, for every $x \in A$ the measure η_x is a Dirac mass supported on the unique trajectory starting from x, and eventually this gives an explicit formula for the solution μ_t, which is therefore unique. □

The importance of the superposition principle also relies in the fact that it will allow, using truncations and restrictions of the measures η_x, several manipulations of solutions of the continuity equation: these constructions are not immediate at the level of the PDE, but they are extremely useful in various occasions, see for instance the proof of Proposition 6.4.3.

This viewpoint is very close in spirit to Young's theory of generalized surfaces and controls [130], a theory with remarkable applications also to nonlinear PDEs (see [83, 122]) and to the calculus of variations (see [28]). There are also strong connections with Brenier's weak solutions of the incompressible Euler equation (see [42]), with Kantorovich's viewpoint in the theory of optimal transportation (see [87, 118]) and with Mather's theory (see [113, 114, 29]).

6.3. The regular Lagrangian flow

As we remarked at the end of Section 6.1, out of the smooth context of the Cauchy–Lipschitz theory the notion of pointwise uniqueness of the ordinary differential equation is not any more the appropriate one. We need to relax the notion of solution, assuming an additional condition which will ensure uniqueness under the various weak differentiability assumptions presented in the previous chapters. With this new concept of solution we will have uniqueness "in the selection sense": every time we approximate the vector field b with a smooth sequence b_h we obtain that the classical flows X_h associated to them converge to the chosen flow for b (see Remark 6.3.3 and Theorem 6.4.6). This notion of solution is encoded in the following definition.

Definition 6.3.1 (Regular Lagrangian flow). Let $b : [0, T] \times \mathbb{R}^d \to \mathbb{R}^d$ be a bounded vector field. We say that a map $X : [0, T] \times \mathbb{R}^d \to \mathbb{R}^d$ is a *regular Lagrangian flow* relative to the vector field b if

(i) for $\mathscr{L}^d$-a.e. $x \in \mathbb{R}^d$ the map $t \mapsto X(t, x)$ is an absolutely continuous integral solution of $\dot{\gamma}(t) = b(t, \gamma(t))$ for $t \in [0, T]$, with $\gamma(0) = x$;

(ii) there exists a constant L independent of t such that

$$X(t, \cdot)_{\#}\mathscr{L}^d \leq L\mathscr{L}^d .$$

The constant L in (ii) will be called the *compressibility constant* of X.

Compare this notion of solution with the one by DiPerna and Lions, presented in Section 6.5. We remark that condition (ii) in Definition 6.3.1 is a *uniform compressibility condition*: we ask that trajectories do not concentrate, and we quantify this with the compressibility constant L. This condition is also important because we obtain a notion of solution which is invariant under modifications of b on negligible sets: if $b(t, x) = \tilde{b}(t, x)$ for $\mathscr{L}^{d+1}$-a.e. $(t, x) \in [0, T] \times \mathbb{R}^d$ it is easy to check that X is a regular Lagrangian flow relative to b if and only if it is a regular Lagrangian flow relative to $\tilde{b}$.

Remark 6.3.2 ($\mathcal{L}_b$-Lagrangian flows). It is also possible to define a more general concept of solution, the one of $\mathcal{L}_b$-Lagrangian flow (see [13, Definition 13]). Consider a convex class $\mathcal{L}_b$ of positive measure-valued solutions μ_t of the continuity equation with vector field b, satisfying the monotonicity assumption

$$0 \leq \mu_t' \leq \mu_t \in \mathcal{L}_b \implies \mu_t' \in \mathcal{L}_b ,$$

whenever μ_t' is a solution of the continuity equation and satisfies an appropriate integrability condition. Then, given a measure $\bar{\mu} \in \mathcal{M}_+(\mathbb{R}^d)$, we can define the notion of $\mathcal{L}_b$-Lagrangian flow starting from $\bar{\mu}$ requiring that the map $X(\cdot, x)$ is an absolutely continuous integral solution of the ordinary differential equation for $\bar{\mu}$-a.e. $x \in \mathbb{R}^d$ and that the measures $\mu_t = X(t, \cdot)_{\#}\bar{\mu}$ induced via push-forward belong to the class $\mathcal{L}_b$. We can see $\mathcal{L}_b$-Lagrangian flows as suitable selections of solutions of the ordinary differential equation, made in such a way to produce densities in $\mathcal{L}_b$. The notion we have given in Definition 6.3.1 corresponds to the case $\mathcal{L}_b = L^{\infty}([0, T] \times \mathbb{R}^d)$. We decided here to focus only on this concept of solution in order to simplify the presentation, but also because tipically the well-posedness results at the PDE level can be shown for bounded solution, hence L^{∞} is the most natural class to be considered (but see also Remark 6.4.2).

Remark 6.3.3. Let us consider again the square root example (see Example 1.2.1):

$$\begin{cases} \dot{\gamma}(t) = \sqrt{|\gamma(t)|} \\ \gamma(0) = x_0. \end{cases}$$

Solutions of the ODE are not unique for $x_0 = -c^2 < 0$. Indeed, they reach the origin in time $2c$, where they can stay for an arbitrary time T, then continuing as $\gamma(t) = \frac{1}{4}(t - T - 2c)^2$. Let us consider for instance the Lipschitz approximation (that could easily be made smooth) of $b(\gamma) = \sqrt{|\gamma|}$ given by

$$b_\epsilon(\gamma) = \begin{cases} \sqrt{|\gamma|} & \text{if } -\infty < \gamma \leq -\epsilon^2 \\ \epsilon & \text{if } -\epsilon^2 \leq \gamma \leq \lambda_\epsilon - \epsilon^2 \\ \sqrt{\gamma - \lambda_\epsilon + 2\epsilon^2} & \text{if } \lambda_\epsilon - \epsilon^2 \leq \gamma < +\infty, \end{cases}$$

with $\lambda_\epsilon - \epsilon^2 > 0$. Then, solutions of the approximating ODEs starting from $-c^2$ reach the value $-\epsilon^2$ in time $t_\epsilon = 2(c - \epsilon)$ and then they continue with constant speed ϵ until they reach $\lambda_\epsilon - \epsilon^2$, in time $T_\epsilon = \lambda_\epsilon/\epsilon$. Then, they continue as $\lambda_\epsilon - 2\epsilon^2 + \frac{1}{4}(t - t_\epsilon - T_\epsilon)^2$.

Choosing $\lambda_\epsilon = \epsilon T$, with $T > 0$, by this approximation we select the solutions that do not move, when at the origin, exactly for a time T.

Other approximations, as for instance $b_\epsilon(\gamma) = \sqrt{\epsilon + |\gamma|}$, select the solutions that move immediately away from the singularity at $\gamma = 0$. Among all possibilities, this family of solutions $\gamma(t, x_0)$ is singled out by the property that $\gamma(t, \cdot)_{\#}\mathscr{L}^1$ is absolutely continuous with respect to $\mathscr{L}^1$, so that no concentration of trajectories occurs at the origin. To see this fact, notice that we can integrate in time the identity

$$0 = \gamma(t, \cdot)_{\#}\mathscr{L}^1(\{0\}) = \mathscr{L}^1(\{x_0 : \gamma(t, x_0) = 0\})$$

and use Fubini's theorem to obtain

$$0 = \int_{\mathbb{R}} \mathscr{L}^1(\{t : \gamma(t, x_0) = 0\}) \, dx_0 \,.$$

Hence, for $\mathscr{L}^1$-a.e. x_0, the trajectory $\gamma(\cdot, x_0)$ does not stay at 0 for a strictly positive set of times.

This example shows that we cannot always expect stability: we have exhibited a family of smooth approximating sequences which converge toward a family of distinct flows of the equation. However, only one of these flows satisfies the compressibility assumption. The point here is the lack of control in L^∞ on the divergence of the vector field (compare with Theorem 6.4.6 and with the first example in Section 5.3).

6.4. Ambrosio's theory of regular Lagrangian flows

In this section we present the derivation of the results of existence, uniqueness and stability for the regular Lagrangian flow deduced from the well-posedness in the class of bounded solutions for the continuity equation. This abstract passage is due to Ambrosio; we refer to [8] for the original approach in the BV case and to the lecture notes [9, 10, 13] for the formalization of the argument in the general case. We notice that these results, in the Sobolev framework, have been already obtained by DiPerna and Lions in [84] with different techniques. We also mention the recent work by Figalli [93] in which a similar theory is developed in the context of stochastic differential equations.

This approach is strongly based on the notion of superposition solution: starting from the "generalized flow" given by the measures $\eta_x \in \mathcal{P}(\mathbb{R}^d)$ we perform various constructions at the level of measure-valued solutions of the continuity equation; at that point the PDE well-posedness comes into play, allowing to deduce results about the measures η_x, roughly speaking obtaining that the generalized flow is in fact a regular Lagrangian flow, since η_x selects a single trajectory for $\mathscr{L}^d$-a.e. $x \in \mathbb{R}^d$.

Theorem 6.4.1. (Existence and uniqueness of the regular Lagrangian flow). *Let $b : [0, T] \times \mathbb{R}^d \to \mathbb{R}^d$ be a bounded vector field. Assume that the continuity equation (6.1) has the uniqueness property in $L^\infty([0, T] \times \mathbb{R}^d)$. Then the regular Lagrangian flow associated to b, if it exists, is unique. Assume in addition that the continuity equation (6.1) with initial data $\bar{\mu} = \mathscr{L}^d$ has a positive solution in $L^\infty([0, T] \times \mathbb{R}^d)$. Then we have existence of a regular Lagrangian flow relative to b.*

With a little abuse of terminology, in the statement of the theorem we write "$\mu_t \in L^\infty$" meaning that we require that μ_t is an absolutely continuous (with respect to $\mathscr{L}^d$) measure, whose density is essentially uniformly bounded. With uniqueness property in $L^\infty([0, T] \times \mathbb{R}^d)$ we mean that, for every initial data $\bar{\mu} \in L^\infty(\mathbb{R}^d)$, if $\mu_t^1, \mu_t^2 \in L^\infty([0, T] \times \mathbb{R}^d)$ are two solutions of (6.1) with $\mu_0^1 = \mu_0^2 = \bar{\mu}$, then we must have $\mu_t^1(x) = \mu_t^2(x)$ for $\mathscr{L}^{d+1}$-a.e. $(t, x) \in [0, T] \times \mathbb{R}^d$.

Notice that, using a simple regularization argument, the existence of a positive solution assumed in the theorem is satified for instance if b has bounded divergence.

Remark 6.4.2 (Uniqueness of $\mathcal{L}_b$-Lagrangian flows). Theorem 6.4.1 could be generalized as follows, recalling the notion of $\mathcal{L}_b$-Lagrangian flow discussed in Remark 6.3.2: if the continuity equation has the uniqueness property in $\mathcal{L}_b$, then there is a unique $\mathcal{L}_b$-Lagrangian flow. The proof is very similar to the one we are going to present: the reader is referred to [13, Section 4] for the details and the modifications needed.

In the remaining part of this section we illustrate the proof of Theorem 6.4.1. We first show that Theorem 6.4.1 is a consequence of the following proposition.

Proposition 6.4.3. *Let* $b : [0, T] \times \mathbb{R}^d \to \mathbb{R}^d$ *be a bounded vector field. Assume that the continuity equation* (6.1) *has the uniqueness property in* $L^\infty([0, T] \times \mathbb{R}^d)$. *Consider a family* $\{\eta_x\}_{x \in \mathbb{R}^d} \subset \mathcal{P}(\Gamma_T)$ *such that* η_x *is concentrated on absolutely continuous integral solutions of the ordinary differential equation starting from* x, *for* $\mathscr{L}^d$-*a.e.* $x \in \mathbb{R}^d$. *Assume that the superposition solution of the continuity equation* $\mu_t^{\eta_x}$ *induced by this family belongs to* $L^\infty([0, T] \times \mathbb{R}^d)$. *Then* η_x *is a Dirac mass for* $\mathscr{L}^d$-*a.e.* $x \in \mathbb{R}^d$.

Remark 6.4.4. We notice that the uniqueness result of Proposition 6.4.3 is in a certain sense stronger than the one of Theorem 6.4.1, since it holds in the wider class of the "multivalued solutions" given by the measures $\{\eta_x\}_{x \in \mathbb{R}^d}$.

Proof of Theorem 6.4.1.
UNIQUENESS OF THE REGULAR LAGRANGIAN FLOW. Assume that there exist two different regular Lagrangian flows $X_1(t, x)$ and $X_2(t, x)$ relative to the vector field b. We show that it is possible to construct a family of measures $\{\eta_x\}_{x \in \mathbb{R}^d} \subset \mathcal{P}(\Gamma_T)$ such that

- for $\mathscr{L}^d$-a.e. $x \in \mathbb{R}^d$ the measure η_x is concentrated on absolutely continuous integral solutions of the ordinary differential equation starting from x;
- the superposition solution $\mu_t^{\eta_x}$ induced by the family $\{\eta_x\}_{x \in \mathbb{R}^d}$ belongs to $L^\infty([0, T] \times \mathbb{R}^d)$;

- the measure η_x is not a Dirac mass for every x belonging to a set with positive Lebesgue measure.

These three properties are clearly in contrast with the result in Proposition 6.4.3: hence, if we are able to construct such a family, we obtain a contradiction that gives the proof of the theorem.

We first consider for $\mathcal{L}^d$-a.e. $x \in \mathbb{R}^d$ the measures $\eta_x^1 = \delta_{X_1(\cdot,x)}$ and $\eta_x^2 = \delta_{X_2(\cdot,x)}$. Define

$$\eta_x = \frac{1}{2}\left(\eta_x^1 + \eta_x^2\right).$$

We check that the superposition solution $\mu_t^{\eta_x}$ is a bounded function. Indeed for every $\varphi \in C_c(\mathbb{R}^d)$ we can compute

$$\langle \mu_t^{\eta_x}, \varphi \rangle = \int_{\mathbb{R}^d}\int_{\Gamma_T} \varphi(\gamma(t))\, d\eta_x(\gamma)\, dx$$

$$= \int_{\mathbb{R}^d}\int_{\Gamma_T} \varphi(\gamma(t))\, d\left(\frac{1}{2}\left(\delta_{X_1(\cdot,x)} + \delta_{X_2(\cdot,x)}\right)\right) dx$$

$$= \int_{\mathbb{R}^d} \frac{1}{2}\left(\varphi(X_1(t,x)) + \varphi(X_2(t,x))\right) dx$$

$$= \frac{1}{2}\int_{\mathbb{R}^d} \varphi(y)\, d\left(X_1(t,\cdot)_\#\mathcal{L}^d\right)(y)$$

$$+ \frac{1}{2}\int_{\mathbb{R}^d} \varphi(y)\, d\left(X_2(t,\cdot)_\#\mathcal{L}^d\right)(y).$$

This means that

$$\mu_t^{\eta_x} = \frac{1}{2}\left(X_1(t,\cdot)_\#\mathcal{L}^d + X_2(t,\cdot)_\#\mathcal{L}^d\right),$$

and by condition (ii) in the definition of regular Lagrangian flow we obtain that $\mu_t^{\eta_x}$ belongs to $L^\infty([0,T]\times\mathbb{R}^d)$. The fact that η_x is concentrated on solutions of the ODE is clear by its definition. Moreover, η_x is not a Dirac mass for every $x \in \mathbb{R}^d$ such that $X_1(\cdot,x) \neq X_2(\cdot,x)$, and this happens in a set of points x of positive measure, since we are assuming that the two regular Lagrangian flows X_1 and X_2 are different.

EXISTENCE OF THE REGULAR LAGRANGIAN FLOW. Let $\mu_t \in L^\infty([0,T]\times\mathbb{R}^d)$ be a positive solution of the continuity equation with initial data $\bar\mu = \mathcal{L}^d$. Noticing that the assumptions of Theorem 6.2.2 are satisfied, we apply the superposition principle deducing that $\mu_t = \mu_t^{\eta_x}$

for some family $\{\eta_x\}_{x\in\mathbb{R}^d}$, with η_x concentrated on absolutely continuous integral solutions of the ODE starting from x, for $\mathscr{L}^d$-a.e. $x \in \mathbb{R}^d$. Since the continuity equation has the uniqueness property in $L^\infty([0, T] \times \mathbb{R}^d)$ we apply Proposition 6.4.3 to deduce that η_x is a Dirac mass for $\mathscr{L}^d$-a.e. $x \in \mathbb{R}^d$. Denote by $X(\cdot, x)$ the element of Γ_T on which η_x is concentrated, for $\mathscr{L}^d$-a.e. $x \in \mathbb{R}^d$; we check that the map $X(t, x)$ defined in this way is a regular Lagrangian flow associated to b. Condition (i) in Definition 6.3.1 is clearly satisfied because η_x is concentrated on solutions of the ODE. To check condition (ii) it is enough to notice that
$$X(t, \cdot)_\# \mathscr{L}^d = \mu_t \in L^\infty([0, T] \times \mathbb{R}^d). \qquad \square$$

We now pass to the proof of Proposition 6.4.3. We will use the following general criterion, whose proof is rather easy.

Lemma 6.4.5. *Let* $\{\eta_x\}_{x\in\mathbb{R}^d} \subset \mathcal{P}(\Gamma_T)$ *satisfy the following property: for every* $t \in [0, T]$ *and every pair of disjoint Borel sets* $E_1, E_2 \subset \mathbb{R}^d$ *we have*

$$\eta_x(\{\gamma : \gamma(t) \in E_1\})\, \eta_x(\{\gamma : \gamma(t) \in E_2\}) = 0 \quad \text{for } \mathscr{L}^d\text{-a.e. } x \in \mathbb{R}^d.$$

Then η_x *is a Dirac mass for* $\mathscr{L}^d$*-a.e,* $x \in \mathbb{R}^d$.

Proof of Proposition 6.4.3. We argue by contradiction and we use the criterion in Lemma 6.4.5. We find $\bar{t} \in]0, T]$, a Borel set $C \subset \mathbb{R}^d$ with $\mathscr{L}^d(C) > 0$ and a couple of disjoint Borel sets $E_1, E_2 \subset \mathbb{R}^d$ such that

$$\eta_x(\{\gamma : \gamma(\bar{t}) \in E_1\})\, \eta_x(\{\gamma : \gamma(\bar{t}) \in E_2\}) \neq 0 \qquad \text{for every } x \in C.$$

Possibly passing to a smaller set C still having strictly positive Lebesgue measure we can assume that

$$0 < \eta_x(\{\gamma : \gamma(\bar{t}) \in E_1\}) \leq M\eta_x(\{\gamma : \gamma(\bar{t}) \in E_2\}) \quad \text{for every } x \in C \tag{6.16}$$

for some constant M. We now want to localize to trajectories starting from the set C and arriving (at time $\bar{t}$) in the sets E_1 and E_2. We define

$$\eta_x^1 = \mathbf{1}_C(x)\eta_x \llcorner \{\gamma : \gamma(\bar{t}) \in E_1\} \quad \text{and} \quad \eta_x^2 = M\mathbf{1}_C(x)\eta_x \llcorner \{\gamma : \gamma(\bar{t}) \in E_2\}.$$

Now denote by μ_t^1 and μ_t^2 (for $t \in [0, \bar{t}]$) the superposition solutions of the continuity equation induced by the families of measures η_x^1 and η_x^2 respectively. Notice that the superposition solutions are well-defined even if η_x^1 and η_x^2 are not probability measures, but just positive measures with finite mass for $\mathscr{L}^d$-a.e. $x \in \mathbb{R}^d$. For every $\varphi \in C_c(\mathbb{R}^d)$ we can compute

$$\langle \mu_0^1, \varphi \rangle = \int_{\mathbb{R}^d} \int_{\Gamma_T} \varphi(\gamma(0)) \, d\Big(\mathbf{1}_C(x) \eta_x \llcorner \{\gamma \,:\, \gamma(\bar{t}) \in E_1\}\Big)(\gamma) \, dx$$

$$= \int_C \int_{\{\gamma \,:\, \gamma(\bar{t}) \in E_1\}} \varphi(\gamma(0)) \, d\eta_x(\gamma) \, dx$$

$$= \int_C \varphi(x) \eta_x\big(\{\gamma \,:\, \gamma(\bar{t}) \in E_1\}\big) \, dx \,,$$

from which we deduce

$$\mu_0^1 = \eta_x\big(\{\gamma \,:\, \gamma(\bar{t}) \in E_1\}\big) \mathscr{L}^d \llcorner C \,.$$

An analogous computation gives

$$\mu_0^2 = M \eta_x\big(\{\gamma \,:\, \gamma(\bar{t}) \in E_2\}\big) \mathscr{L}^d \llcorner C \,.$$

Recalling (6.16) we obtain that $\mu_0^1 \le \mu_0^2$. Now let $f : \mathbb{R}^d \to [0, 1]$ be the density of μ_0^1 with respect to μ_0^2 (i.e. f satisfies $\mu_0^1 = f\mu_0^2$) and set

$$\tilde{\eta}_x^2 = M f(x) \mathbf{1}_C(x) \eta_x \llcorner \{\gamma \,:\, \gamma(\bar{t}) \in E_2\} \,.$$

Consider the superposition solution $\tilde{\mu}_t^2$ (defined for $t \in [0, \bar{t}]$) induced by the family of measures $\tilde{\eta}_x^2$. We can readily check that $\mu_0^1 = \tilde{\mu}_0^2$ and that

$$\langle \mu_t^1, \varphi \rangle = \int_C \int_{\{\gamma \,:\, \gamma(\bar{t}) \in E_1\}} \varphi(\gamma(\bar{t})) \, d\eta_x(\gamma) \, dx$$

and

$$\langle \tilde{\mu}_t^2, \varphi \rangle = \int_C M f(x) \int_{\{\gamma \,:\, \gamma(\bar{t}) \in E_2\}} \varphi(\gamma(\bar{t})) \, d\eta_x(\gamma) \, dx \,.$$

We deduce that μ_t^1 is concentrated on E_1 and $\tilde{\mu}_t^2$ is concentrated on E_2. Hence μ_t^1 and $\tilde{\mu}_t^2$ are solutions in $L^\infty([0, T] \times \mathbb{R}^d)$ of the continuity equation with the same initial data at time $t = 0$, but they are different at time $t = \bar{t}$. We are violating the uniqueness assumption and from this contradiction we obtain the thesis. $\square$

In a similar fashion it is possible to show a result of abstract stability: if the continuity equation is well-posed for each approximating vector field b_k then the associated regular Lagrangian flows converge strongly to the regular Lagrangian flow associated to the limit vector field b.

Theorem 6.4.6 (Stability of the regular Lagrangian flow). *Consider a sequence* $\{b_k\}$ *of vector fields such that*

$$\|b_k\|_{L^\infty([0,T]\times\mathbb{R}^d)} + \|\operatorname{div} b_k\|_{L^\infty([0,T]\times\mathbb{R}^d)} \le C \qquad (6.17)$$

and assume that for each b_k *the continuity equation has the uniqueness property in* $L^\infty([0,T]\times\mathbb{R}^d)$. *Assume that the sequence* $\{b_k\}$ *converges in* $L^1_{\text{loc}}([0,T]\times\mathbb{R}^d)$ *to a vector field* $b \in L^\infty([0,T]\times\mathbb{R}^d)$ *with* $\operatorname{div} b \in L^\infty([0,T]\times\mathbb{R}^d)$. *Assume that the continuity equation with vector field* b *has the uniqueness property in* $L^\infty([0,T]\times\mathbb{R}^d)$. *Then the regular Lagrangian flows* X_k *associated to* b_k *converge strongly in* $L^\infty([0,T]; L^1_{\text{loc}}(\mathbb{R}^d))$ *to the regular Lagrangian flow* X *associated to* b.

We only give a sketch of the proof, which is closely related to the one of the uniqueness result of Theorem 6.4.1. We refer the reader to [10, Section 3] for more details.

As in the proof of the superposition principle we consider for every k the measure $\eta^k \in \mathcal{M}_+(\mathbb{R}^d \times \Gamma_T)$ associated to the regular Lagrangian flow X^k. Using the bounds (6.17) it is possible to show that η^k is locally tight. Consider a limit point $\eta \in \mathcal{M}_+(\mathbb{R}^d \times \Gamma_T)$. Arguing as in Step 6 of the proof of Theorem 6.2.2 and using the convergence of b_k to b we obtain that η is concentrated on trajectories of the ODE with vector field b. Moreover from the narrow convergence and from (6.17) we get that η induces bounded superposition solutions to the continuity equation. We can then apply Proposition 6.4.3 to get that η is in fact concentrated on a graph of a regular Lagrangian flow associated to b. But under our assumptions we already know from Theorem 6.4.1 that such a regular Lagrangian flow is unique, and this means that we have

$$(x, X^k(\cdot, x))_\#\mathscr{L}^d \to (x, X(\cdot, x))_\#\mathscr{L}^d$$

locally narrowly. From this it is possible to deduce the strong convergence of X^k to X in $L^\infty([0,T]; L^1_{\text{loc}}(\mathbb{R}^d))$.

Remark 6.4.7. The strength of this abstract approach relies in the fact that, having at our disposal the PDE well-posedness results for Sobolev or BV vector fields (see Sections 2.5 and 2.6), for the various cases addressed in Section 2.7 and for the two-dimensional case (see Chapter 4) we can immediately conclude, using Theorems 6.4.1 and 6.4.6, existence, uniqueness and stability of the regular Lagrangian flow in all these situations (see also Remarks 2.3.5 and 2.3.6).

6.5. DiPerna–Lions' notion of flow

The approach to the ordinary differential equation due to DiPerna and Lions [84] is quite different. It is based on an observation that we already made in the smooth setting: the flow $X(t, s, x)$ starting at time s from the point x satisfies

$$\frac{\partial X}{\partial s}(t, s, x) + \big(b(s, x) \cdot \nabla_x\big) X(t, s, x) = 0$$

(see equation (1.15)), that is a transport equation with vector field b. In this way results relative to the PDE can be transferred to the ODE. DiPerna–Lions' notion of flow is based on the following three axioms:

(a) The ordinary differential equation

$$\begin{cases} \dfrac{\partial}{\partial t} X(t, x) = b(t, X(t, x)) \\[2mm] X(0, x) = x \end{cases}$$

is satisfied in the sense of distributions in $[0, T] \times \mathbb{R}^d$;
(b) The push-forward of the Lebesgue measure satisfies

$$\frac{1}{C} \mathscr{L}^d \leq X(t, \cdot)_\# \mathscr{L}^d \leq C \mathscr{L}^d;$$

(c) The semi-group property holds: for $\mathscr{L}^d$-a.e. $x \in \mathbb{R}^d$ we have

$$X(t, X(s, x)) = X(s + t, x) \qquad \text{for every } s, t \geq 0 \text{ with } s + t \leq T.$$

Up to a redefinition of the flow X on a negligible set, condition (a) is equivalent to condition (i) in the definition of regular Lagrangian flow. However, Definition 6.3.1(ii) only asks for an upper bound on the push-forward of the Lebesgue measure, while (b) also requires a bound from below. Moreover, the semi-group property (c) is completely removed in the approach of [8]: it is in fact a consequence of the other two assumptions, using the following argument (see [8, Remark 6.7]).
    Fix $s \in [s', T]$ and define

$$\tilde{X}(t, x) = \begin{cases} X(t, s', x) & \text{if } t \in [s', s] \\ X(t, s, X(s, s', x)) & \text{if } t \in [s, T]. \end{cases}$$

It is immediate to check that $\tilde{X}(\cdot, x)$ is a solution of the ODE in $[s', T]$ for $\mathscr{L}^d$-a.e. $x \in \mathbb{R}^d$. Moreover we easily obtain the following compressibility condition for $\tilde{X}$:

$$\tilde{X}(t, \cdot)_\# \mathscr{L}^d \leq L^2 \mathscr{L}^d,$$

where L is a compressibility constant for the regular Lagrangian flow X. By the uniqueness result we obtain that $\tilde{X}(t, x) = X(t, s', x)$, and this means that the semi-group property (c) holds.

In this framework we also mention the recent result by Hauray, Le Bris and Lions [101] in which uniqueness of this notion of flow (for Sobolev vector fields) is shown directly at the ODE level, with a very simple argument.

6.6. The density of a regular Lagrangian flow

We now present another approach to the theory of regular Lagrangian flows, based on the concept of density transported by the flow and particularly useful to deal with nearly incompressible vector fields (recall Definition 2.8.1); see Section 2.8 for the PDE theory in this framework. The presentation here is very close to the one by De Lellis ([77, 79]) and we refer to these papers for a complete exposition.

This approach works naturally under a compressibility assumption which is a bit weaker than the one assumed in the standard definition of regular Lagrangian flow (see Definition 6.3.1(ii)): we will only need that

$$X(t, \cdot)_{\#}\mathscr{L}^d \ll \mathscr{L}^d \qquad \text{for } \mathscr{L}^1\text{-a.e. } t \in [0, T]. \tag{6.18}$$

If X is a regular Lagrangian flow associated to a bounded vector field b we define

$$\mu_X = (\mathrm{id}, X)_{\#}\big(\mathscr{L}^{d+1} \llcorner ([0, T] \times \mathbb{R}^d)\big),$$

i.e. μ_X is the push-forward of the Lebesgue measure on $[0, T] \times \mathbb{R}^d$ via the map $(t, x) \mapsto (t, X(t, x))$. The compressibility condition (6.18) gives the existence of a function $\rho \in L^1_{\mathrm{loc}}([0, T] \times \mathbb{R}^d)$ such that

$$\mu_X = \rho\mathscr{L}^{d+1} \llcorner ([0, T] \times \mathbb{R}^d). \tag{6.19}$$

As already remarked in Section 2.8, in the case of a smooth vector field we can explicitly compute

$$\rho(t, x) = \det \nabla_x X\big(t, X(t, \cdot)^{-1}(x)\big).$$

Definition 6.6.1. The function ρ defined via (6.19) is called the *density of the regular Lagrangian flow* X.

The following proposition establishes a link between this notion and the Eulerian side of the problem, namely with the continuity equation relative to the vector field b.

Proposition 6.6.2. *Let X be a regular Lagrangian flow associated to a bounded nearly incompressible vector field b and let $\bar{\zeta} \in L^\infty(\mathbb{R}^d)$. Define*

$$\mu = \big(\mathrm{id}, X\big)_{\#}\big(\bar{\zeta}\, \mathscr{L}^{d+1} \llcorner ([0, T] \times \mathbb{R}^d)\big).$$

Then there exists $\zeta \in L^1_{\mathrm{loc}}([0, T] \times \mathbb{R}^d)$ such that $\mu = \zeta \mathscr{L}^{d+1} \llcorner ([0, T] \times \mathbb{R}^d)$; moreover ζ is a solution of

$$\begin{cases} \partial_t \zeta + \mathrm{div}\,(b\zeta) = 0 \\ \zeta(0, \cdot) = \bar{\zeta}. \end{cases} \tag{6.20}$$

The density of the regular Lagrangian flow corresponds to the case $\bar{\zeta} = 1$ in the previous proposition: this means that the density ρ satisfies

$$\begin{cases} \partial_t \rho + \mathrm{div}\,(b\rho) = 0 \\ \rho(0, \cdot) = 1. \end{cases}$$

In particular we deduce that, when b is a bounded nearly incompressible vector field satisfying the renormalization property (recall Definition 2.8.2), the density of a regular Lagrangian flow X relative to b coincides with the density generated by b in the sense of Definition 2.8.5. This is precisely the link between the Lagrangian and the Eulerian problems: the proof of the following theorem, relative to the well-posedness for the regular Lagrangian flow, is strongly based on the well-posedness for the continuity equation in Theorem 2.8.4.

Theorem 6.6.3. *Assume that b is a bounded nearly incompressible vector field such that the extension $\tilde{b}$ as in (2.36) has the renormalization property in the sense of Definition 2.8.2. Then there exists a unique regular Lagrangian flow X associated to b.*

Moreover, let $\{b_k\}$ be a sequence of bounded nearly incompressible vector fields such that the extensions $\tilde{b}_k$ as in (2.36) have the renormalization property in the sense of Definition 2.8.2 and, for every k, let ρ_k be as in (2.35). Assume that there exists a constant C such that

$$\|b_k\|_\infty + \|\rho_k\|_\infty + \|\rho_k^{-1}\|_\infty \le C$$

and that $b_k \to b$ in $L^1_{\mathrm{loc}}([0, T] \times \mathbb{R}^d)$. Then the regular Lagrangian flows X_k associated to b_k converge strongly in $L^1_{\mathrm{loc}}([0, T] \times \mathbb{R}^d)$ to X.

Proof. The existence part is essentially based on an approximation procedure, which however requires some care, and for this we refer to [77, Theorem 3.22].

We now show uniqueness. Let X and Y be two regular Lagrangian flows relative to the vector field b. Fix $\bar{\zeta} \in C_c(\mathbb{R}^d)$ and, recalling Theorem 2.8.4, consider the unique solution ζ of (6.20). According to Proposition 6.6.2 we have

$$\zeta \mathscr{L}^{d+1} \llcorner \big([0, T] \times \mathbb{R}^d\big) = (\mathrm{id}, X)_{\#}\big(\bar{\zeta}\mathscr{L}^{d+1}\llcorner \big([0, T] \times \mathbb{R}^d\big)\big)$$
$$= (\mathrm{id}, Y)_{\#}\big(\bar{\zeta}\mathscr{L}^{d+1}\llcorner \big([0, T] \times \mathbb{R}^d\big)\big).$$

This identity means

$$\int_0^T \int_{\mathbb{R}^d} \varphi(x, X(t, x))\bar{\zeta}(x)dx\,dt = \int_0^T \int_{\mathbb{R}^d} \varphi(x, Y(t, x))\bar{\zeta}(x)dx\,dt$$

for every $\varphi \in C_c([0, T] \times \mathbb{R}^d)$. Since $\bar{\zeta}$ has compact support this identity is also valid when $\varphi(t, y) = \chi(t)y_j$, for $\chi \in C([0, T])$ and $j = 1, \ldots, d$. Then

$$\int_0^T \int_{\mathbb{R}^d} X_j(t, x)\chi(t)\bar{\zeta}(x)\,dx\,dt = \int_0^T \int_{\mathbb{R}^d} Y_j(t, x)\chi(t)\bar{\zeta}(x)\,dx\,dt$$

for any pair of functions $\chi \in C([0, T])$ and $\bar{\zeta} \in C_c(\mathbb{R}^d)$. This implies that $X(t, x) = Y(t, x)$ for $\mathscr{L}^{d+1}$-a.e. $(t, x) \in [0, T] \times \mathbb{R}^d$.

We finally show the stability property. Let $b_k \to b$ as in the statement of the theorem. Fix a function $\bar{\zeta} \in C_c(\mathbb{R}^d)$. Applying Theorem 2.8.4 we can define ζ and ζ_k as the unique solutions to the problems

$$\begin{cases} \partial_t \zeta + \mathrm{div}\,(b\zeta) = 0 \\ \zeta(0, \cdot) = \bar{\zeta} \end{cases} \quad \text{and} \quad \begin{cases} \partial_t \zeta_k + \mathrm{div}\,(b_k\zeta_k) = 0 \\ \zeta_k(0, \cdot) = \bar{\zeta}. \end{cases}$$

By the weak stability of the continuity equation we deduce that, up to subsequences, $\zeta_k \overset{*}{\rightharpoonup} \zeta$ in $L^\infty([0, T] \times \mathbb{R}^d) - w^*$. Recalling the characterization in Proposition 6.6.2 we see that this means

$$\int_0^T \int_{\mathbb{R}^d} \varphi\big(t, X_k(t, x)\big)\bar{\zeta}(x)dt\,dx \to \int_0^T \int_{\mathbb{R}^d} \varphi\big(t, X(t, x)\big)\bar{\zeta}(x)dt\,dx \quad (6.21)$$

for every $\varphi \in C_c([0, T] \times \mathbb{R}^d)$. However, noticing that the regular Lagrangian flows X_k are locally equi-bounded, we are allowed to use as test functions $\bar{\zeta}(x) = \mathbf{1}_{B_R(0)}(x)$ and $\varphi(t, x) = |x|^2$, and this gives

$$\int_0^T \int_{B_R(0)} |X_k(t, x)|^2\,dt\,dx \to \int_0^T \int_{B_R(0)} |X(t, x)|^2\,dt\,dx. \quad (6.22)$$

For the same reason we can substitute in (6.21) $\bar{\zeta}(x) = \beta(x)\mathbf{1}_{B_R(0)}(x)$ and $\varphi(t, x) = \gamma(t)x \cdot v$, where $\beta \in C_c(\mathbb{R}^d)$, $\gamma \in C([0, T])$ and $v \in \mathbb{R}^d$. Since we can approximate X strongly in $L^1_{\text{loc}}([0, T] \times \mathbb{R}^d)$ with functions of the form

$$\sum_{i=1}^{N} v_i \gamma_i(t)\beta_i(x)$$

we obtain

$$\int_0^T \int_{B_R(0)} X_k(t, x) \cdot X(t, x)\, dt dx \to \int_0^T \int_{B_R(0)} |X(t, x)|^2\, dt dx \,. \quad (6.23)$$

Putting together (6.22) and (6.23) we get the desired conclusion.   □

6.7. Bressan's compactness conjecture

In [48] Bressan proposed the following conjecture.

Conjecture 6.7.1 (Bressan's compactness conjecture). Let $b_k : [0, T] \times \mathbb{R}^d \to \mathbb{R}^d$, $k \in \mathbb{N}$, be a sequence of smooth vector fields and denote by X_k the (classical) flows associated to them, i.e. the solutions of

$$\begin{cases} \dfrac{\partial X_k}{\partial t}(t, x) = b_k(t, X_k(t, x)) \\[2mm] X_k(0, x) = x\,. \end{cases} \quad (6.24)$$

Assume that $\|b_k\|_\infty + \|\nabla b_k\|_{L^1}$ is uniformly bounded and that the flows X_k are uniformly nearly incompressible, i.e. that

$$\frac{1}{C} \le \det(\nabla_x X_k(t, x)) \le C \qquad \text{for some constant } C > 0. \quad (6.25)$$

Then the sequence $\{X_k\}$ is strongly precompact in $L^1_{\text{loc}}([0, T] \times \mathbb{R}^d)$.

This conjecture was advanced in connection with the Keyfitz and Kranzer system, in particular to provide the existence of suitable weak solutions (recall Section 5.2). In fact it is possible to prove well-posedness results for this system bypassing Conjecture 6.7.1 (see [15, 11]), which nevertheless remains an interesting question. In this section we indicate some advances in its proof, mainly contained in [16].

In [11] it has been proved that Conjecture 6.7.1 would follow from the following one.

Conjecture 6.7.2. Any bounded nearly incompressible BV vector field has the renormalization property in the sense of Definition 2.8.2.

Conjecture 6.7.2 *implies Conjecture* 6.7.1. Let ρ_k be the density generated by X_k. From (6.25) it follows the existence of a constant $\tilde{C} > 0$ such that

$$\frac{1}{\tilde{C}} \leq \rho_k \leq \tilde{C}.$$

From the BV compactness theorem and the weak* compactness of L^∞ it is sufficient to prove Conjecture 6.7.1 under the assumptions that $b_k \to b$ strongly in $L^1_{\mathrm{loc}}([0, T] \times \mathbb{R}^d)$ for some BV vector field b and that $\rho_k \overset{*}{\rightharpoonup} \rho$ in $L^\infty([0, T] \times \mathbb{R}^d) - w^*$ for some bounded function ρ. We notice that this implies that b is bounded, $\rho \geq 1/\tilde{C}$ and

$$\partial_t \rho + \mathrm{div}\,(b\rho) = 0.$$

The last identity can be deduced passing to the limit in the equations $\partial_t \rho_k + \mathrm{div}\,(b_k \rho_k) = 0$.

Then b is a bounded nearly incompressible BV vector field and, if Conjecture 6.7.2 has a positive answer, b has the renormalization property. If this is the case, we apply the stability result in Theorem 6.6.3 to conclude that X_k converges strongly in $L^1_{\mathrm{loc}}([0, T] \times \mathbb{R}^d)$ to the unique regular Lagrangian flow associated to b. $\qquad\square$

In order to present the main results of [16] we need to introduce some notation, taken from [14]. Let $\Omega \subset \mathbb{R}^m$ be an open set and consider $B \in BV_{\mathrm{loc}}(\Omega; \mathbb{R}^m)$. Consider the set $\tilde{E}$ consisting of the points $x \in \Omega$ such that the limit

$$M(x) = \lim_{r \to 0} \frac{DB(B_r(x))}{|DB|(B_r(x))}$$

exists and is finite and the Lebesgue limit $\tilde{B}(x)$ exists. The tangential set of B is defined as follows:

$$E = \left\{ x \in \tilde{E} \; : \; M(x) \cdot \tilde{B}(x) = 0 \right\}. \tag{6.26}$$

We can think of this set as the set on which the vector field B is "orthogonal to its derivative". Building on some results of [14], in [16] the following result is shown.

Theorem 6.7.3. *Let* $b : [0, T] \times \mathbb{R}^d \to \mathbb{R}^d$ *be a bounded nearly incompressible* BV *vector field. Consider the vector field* $B = (1, b) :$ $[0, T] \times \mathbb{R}^d \to \mathbb{R} \times \mathbb{R}^d$ *and let* E *be the tangential set of* B. *If*

$$|D^c_{t,x} \cdot B|(E) = |D^c_x \cdot b|(E) = 0 \tag{6.27}$$

then b *has the renormalization property.*

In the statement of the theorem we used the notation $D^c_{t,x} \cdot B$ and $D^c_x \cdot b$ for the Cantor part of the space-time and of the space divergence of B and b respectively. The presence of condition (6.27) in the theorem leads to the following question.

Question 6.7.4. Let $B \in BV_{\text{loc}} \cap L^\infty_{\text{loc}}(\Omega; \mathbb{R}^m)$ and denote by E its tangential set. Under which conditions does the equality $|D^c \cdot B|(E) = 0$ hold?

Indeed some conditions are needed, as shown by a counterexample presented in [16, Section 8]. Notice that a positive answer to the following question would be enough for the proof of Conjecture 6.7.2; however, the answer to this second question is presently not known.

Question 6.7.5. Let $B \in BV_{\text{loc}} \cap L^\infty_{\text{loc}}(\Omega; \mathbb{R}^m)$ and denote by E its tangential set. Assume that there exists a function $\rho \in L^\infty(\Omega)$ such that $\rho \geq C > 0$ and div $(B\rho) = 0$. Then does the equality $|D^c \cdot B|(E) = 0$ hold?

Another approach to Bressan's compactness conjecture, based on quantitative a priori estimates for regular Lagrangian flows (and hence bypassing the connection with the theory of renormalized solutions), will be presented in the next chapter.

Chapter 7
A priori estimates for regular Lagrangian flows

In this chapter we present a joint work with De Lellis [66] (see also [67] and [13, Section 8] for another exposition of these results). We are able to show some quantitative estimates for $W^{1,p}$ vector fields, with $p > 1$, which allow to recover the results of existence, uniqueness and stability of regular Lagrangian flows presented in the previous chapter. Moreover these estimates have some new and interesting corollaries, regarding compactness, quantitative regularity and quantitative stability of regular Lagrangian flows. They also allow to prove a propagation of a mild regularity for solutions to the transport equation. One of the merits of this approach is the fact that the whole derivation is purely Lagrangian, in the sense that everything is deduced from the estimates and from the definition of regular Lagrangian flow only, with no mention to the Eulerian side of the problem (compare with the derivation we presented in Section 6.4). The only drawback lies in the assumption $p > 1$, which can be relaxed a bit (in fact our arguments work under the assumption $Db \in L \log L$, see Sections 7.2 and 7.4), but we are presently not able to cover the case $p = 1$, and this does not allow to reach the very important BV setting.

7.1. A purely Lagrangian approach

We recall that, as we discussed in Section 6.3, for a vector field b which is merely locally summable we can give the following definition of *regular Lagrangian flow*. This notion turns out to be the right one in the study of the ordinary differential equation with weakly differentiable vector field. Notice that the boundedness assumption on the vector field assumed in the previous chapters is not essential: in this chapter we will also address the theory of regular Lagrangian flows relative to vector fields satisfying more general growth conditions.

Definition 7.1.1 (Regular Lagrangian flow). Let $b \in L^1_{\text{loc}}([0, T] \times \mathbb{R}^d; \mathbb{R}^d)$. We say that a map $X : [0, T] \times \mathbb{R}^d \to \mathbb{R}^d$ is a *regular Lagrangian flow* for the vector field b if

(i) for $\mathscr{L}^d$-a.e. $x \in \mathbb{R}^d$ the map $t \mapsto X(t, x)$ is an absolutely continuous
     integral solution of $\dot{\gamma}(t) = b(t, \gamma(t))$ for $t \in [0, T]$, with $\gamma(0) = x$;
(ii) there exists a constant L independent of t such that

$$X(t, \cdot)_\# \mathscr{L}^d \leq L \mathscr{L}^d . \tag{7.1}$$

The constant L in (ii) will be called the *compressibility constant* of X.

   We illustrated in the previous chapter the results by DiPerna and Lions
[84] and by Ambrosio [8] regarding existence, uniqueness and stability
of regular Lagrangian flows, in the Sobolev and BV context respectively.
However the argument that we have shown is quite indirect: as we have
seen it exploits the connection between the ordinary differential equation
and the Cauchy problem for the continuity equation.

   We show in this chapter how many of the results of the DiPerna–Lions
theory relative to regular Lagrangian flows can be recovered from simple a priori estimates, directly in the Lagrangian formulation. Though
our approach works under various relaxed hypotheses, namely controlled
growth at infinity of the field b and L^p_{loc} and $L \log L$ assumptions on $D_x b$,
for simplicity in this introductory discussion we consider a vector field b
in $W^{1,p} \cap L^\infty$, with $p > 1$. Assuming the existence of a regular Lagrangian flow X, we give estimates of integral quantities depending on
$X(t, x) - X(t, y)$. These estimates depend only on $\|b\|_{W^{1,p}} + \|b\|_\infty$ and
the compressibility constant L of Definition 7.1.1(ii). Moreover, a similar
estimate can be derived for the difference $X(t, x) - X'(t, x)$ of regular Lagrangian flows of different vector fields b and b', depending only on the
compressibility constant of b and on $\|b\|_{W^{1,p}} + \|b\|_\infty + \|b'\|_\infty + \|b - b'\|_{L^1}$.
As direct corollaries of our estimates we then derive:

(a) Existence, uniqueness, stability, and compactness of regular Lagran-
    gian flows;
(b) Some mild regularity properties, like the approximate differentiability
    proved in [19], that we recover in a new quantitative fashion.

The regularity property in (b) has an effect on solutions to the transport
equation: we can prove that, for $b \in W^{1,p} \cap L^\infty$ with bounded divergence,
solutions to the transport equation propagate the same mild regularity of
the corresponding regular Lagrangian flow (we refer to Section 7.5 for
the precise statements).

   Our approach has been inspired by a recent result of Ambrosio,
Lecumberry and Maniglia [19], proving the almost everywhere approximate differentiability of regular Lagrangian flows. Indeed, some of the
quantities we estimate in this paper are taken directly from [19], whereas
others are just suitable modifications. However, the way we derive our

estimates is different: our analysis relies all on the Lagrangian formulation, whereas that of [19] relies on the Eulerian one. We also mention a previous work by Le Bris and Lions [104], in which, among other things, a kind of "differentiability in measure" of the flow is shown. The relation between this notion and the classical approximate differentiability has been studied by Ambrosio and Malý [21]. See also [13, Sections 6 and 7] for an account of these results.

Unfortunately with our approach we do not recover all the results of the theory of renormalized solutions. The main problem is that our estimates do not conver the case $Db \in L^1$. Actually, the extension to the case $Db \in L^1$ of our (or of similar) estimates would answer positively to the compactness conjecture made by Bressan in [48] (see Conjecture 6.7.1). At the present stage, the theory of renormalized solutions cannot be extended to cover this interesting case (see Section 6.7 for an account of some partial results in this direction). In another paper, [47], Bressan raised a second conjecture on mixing properties of flows of BV vector fields (see Conjecture 7.6.1 below), which can be considered as a quantitative version of Conjecture 6.7.1. In Section 6 we show how our estimates settle the $W^{1,p}$ ($p > 1$) analog of Bressan's mixing conjecture.

In order to keep the presentation simple, in Section 7.2 we give the estimates and the various corollaries in the case $b \in W^{1,p} \cap L^\infty$ and in Section 7.3 we present the more general estimates and their consequences. We thank Herbert Koch for suggesting us that the Lipschitz estimates hold under the assumption $Db \in L \log L$ (see Remark 7.2.4 and the discussion at the beginning of Section 7.4). In Section 7.4 we show how to prove directly, via suitable a priori estimates, the compactness conclusion of Conjecture 6.7.1 when Db_k is bounded in $L \log L$. It has been pointed out independently by François Bouchut and by Pierre-Emmanuel Jabin that a more careful analysis allows to extend this approach when the sequence $\{Db_k\}$ is equi-integrable. In Section 7.5 we discuss the regularity results for transport equations mentioned above. Finally, in Section 7.6 we prove the $W^{1,p}$ analog of Bressan's mixing conjecture.

In all the chapter constants will be denoted by c and $c_{a_1,\dots,a_q}$, where we understand that in the first case the constant is universal and in the latter that it depends only on the quantities $a_1, \dots, a_q$. Therefore, during several computations, we will use the same symbol for constants which change from line to line.

7.2. A priori estimates for bounded vector fields and corollaries

In this section we show our estimates in the particular case of bounded vector fields. This estimate and its consequences are just particular cases of the more general theorems presented in the next sections. However, we decided to give independent proofs in this simplified setting in order to illustrate better the basic ideas of our analysis.

7.2.1. Estimate of an integral quantity and Lipschitz estimates

Theorem 7.2.1. *Let b be a bounded vector field belonging to $L^1([0, T];$ $W^{1,p}(\mathbb{R}^d; \mathbb{R}^d))$ for some $p > 1$ and let X be a regular Lagrangian flow associated to b. Let L be the compressibility constant of X, as in Definition 7.1.1(ii). For every $p > 1$ define the following integral quantity:*

$$A_p(R, X)$$
$$= \left[\int_{B_R(0)} \left(\sup_{0 \le t \le T} \sup_{0 < r < 2R} \fint_{B_r(x)} \log\left(\frac{|X(t, x) - X(t, y)|}{r} + 1 \right) dy \right)^p dx \right]^{1/p}.$$

Then we have

$$A_p(R, X) \le C\left(R, L, \|D_x b\|_{L^1(L^p)}\right). \tag{7.2}$$

Remark 7.2.2. A small variant of the quantity $A_1(R, X)$ was first introduced in [19] and studied in an Eulerian setting in order to prove the approximate differentiability of regular Lagrangian flows. One basic observation of [19] is that a control of $A_1(R, X)$ implies the Lipschitz regularity of X outside of a set of small measure. This elementary Lipschitz estimate is shown in Proposition 7.2.3. The novelty of our point of view is that a direct Lagrangian approach allows to derive uniform estimates as in (7.2). These uniform estimates are then exploited in the next subsections to show existence, uniqueness, stability and regularity of the regular Lagrangian flow.

All the computations in the following proof can be justified using the definition of regular Lagrangian flow (Definition 7.1.1): the differentiation of the flow with respect to the time gives the vector field (computed along the flow itself), thanks to condition (i); condition (ii) implies that, when we perform a change of variable, we can estimate the result from above just multiplying by the compressibility constant L.

During the proof, we will use some tools borrowed from the theory of maximal functions. We recall that, for a function $f \in L^1_{\text{loc}}(\mathbb{R}^d; \mathbb{R}^m)$, the *local maximal function* is defined as

$$M_\lambda f(x) = \sup_{0<r<\lambda} \fint_{B_r(x)} |f(y)| \, dy \, .$$

For more details about the maximal function and for the statements of the lemmas we are going to use, we refer to Appendix A.4.

Proof of Theorem 7.2.1. For $0 \le t \le T$, $0 < r < 2R$ and $x \in B_R(0)$ define

$$Q(t, x, r) = \fint_{B_r(x)} \log\left(\frac{|X(t, x) - X(t, y)|}{r} + 1\right) dy \, .$$

From Definition 7.1.1(i) it follows that for $\mathscr{L}^d$-a.e. x and for every $r > 0$ the map $t \mapsto Q(t, x, r)$ is Lipschitz and

$$\frac{dQ}{dt}(t,x,r) \le \fint_{B_r(x)} \left| \frac{dX}{dt}(t,x) - \frac{dX}{dt}(t,y) \right| (|X(t,x) - X(t,y)| + r)^{-1} \, dy$$

$$= \fint_{B_r(x)} \frac{|b(t, X(t, x)) - b(t, X(t, y))|}{|X(t, x) - X(t, y)| + r} dy \, .$$

(7.3)

We now set $\tilde{R} = 4R + 2T\|b\|_\infty$. Since we clearly have $|X(t, x) - X(t, y)| \le \tilde{R}$, applying Lemma A.4.3 we can estimate

$$\frac{dQ}{dt}(t, x, r) \le c_d \fint_{B_r(x)} \left(M_{\tilde{R}} Db(t, X(t, x)) \right.$$

$$\left. + M_{\tilde{R}} Db(t, X(t, y)) \right) \frac{|X(t, x) - X(t, y)|}{|X(t, x) - X(t, y)| + r} \, dy$$

$$\le c_d M_{\tilde{R}} Db(t, X(t, x)) + c_d \fint_{B_r(x)} M_{\tilde{R}} Db(t, X(t, y)) \, dy \, .$$

(7.4)

Integrating with respect to the time, passing to the supremum for $0 < r < 2R$ and exchanging the supremums we obtain

$$\sup_{0 \le t \le T} \sup_{0 < r < 2R} Q(t, x, r) \le c + c_d \int_0^T M_{\tilde{R}} Db(t, X(t, x)) \, dt$$

(7.5)

$$+ c_d \int_0^T \sup_{0 < r < 2R} \fint_{B_r(x)} M_{\tilde{R}} Db(t, X(t, y)) \, dy dt \, .$$

Taking the L^p norm over $B_R(0)$ we get

$$A_p(R, X) \le c_{p,R} + c_d \left\| \int_0^T M_{\tilde{R}} Db(t, X(t, x)) \, dt \right\|_{L^p(B_R(0))} \tag{7.6}$$

$$+ c_d \left\| \int_0^T \sup_{0 < r < 2R} \fint_{B_r(x)} M_{\tilde{R}} Db(t, X(t, y)) \, dy dt \right\|_{L^p(B_R(0))} . \tag{7.7}$$

Recalling Definition 7.1.1(ii) and Lemma A.4.2, the integral in (7.6) can be estimated with

$$c_d L^{1/p} \int_0^T \left\| M_{\tilde{R}} Db(t, x) \right\|_{L^p(B_{R+T\|b\|_\infty}(0))} \, dt$$
$$\le c_{d,p} L^{1/p} \int_0^T \left\| Db(t, x) \right\|_{L^p(B_{R+\tilde{R}+T\|b\|_\infty}(0))} \, dt . \tag{7.8}$$

The integral in (7.7) can be estimated in a similar way with

$$c_d \int_0^T \left\| \sup_{0 < r < 2R} \fint_{B_r(x)} \left[(M_{\tilde{R}} Db) \circ (t, X(t, \cdot)) \right](y) \, dy \right\|_{L^p(B_R(0))} \, dt$$

$$= c_d \int_0^T \left\| M_{2R} \left[(M_{\tilde{R}} Db) \circ (t, X(t, \cdot)) \right](x) \right\|_{L^p(B_R(0))} \, dt$$

$$\le c_{d,p} \int_0^T \left\| \left[(M_{\tilde{R}} Db) \circ (t, X(t, \cdot)) \right](x) \right\|_{L^p(B_{3R}(0))} \, dt$$

$$= c_{d,p} \int_0^T \left\| (M_{\tilde{R}} Db) \circ (t, X(t, x)) \right\|_{L^p(B_{3R}(0))} \, dt$$

$$\le c_{d,p} L^{1/p} \int_0^T \left\| M_{\tilde{R}} Db(t, x) \right\|_{L^p(B_{3R+T\|b\|_\infty}(0))} \, dt$$

$$\le c_{d,p} L^{1/p} \int_0^T \left\| Db(t, x) \right\|_{L^p(B_{3R+T\|b\|_\infty+\tilde{R}}(0))} \, dt . \tag{7.9}$$

Combining (7.6), (7.7), (7.8) and (7.9), we obtain the desired estimate for $A_p(R, X)$. □

We now show how the estimate of the integral quantity gives a quantitative Lipschitz estimate.

Proposition 7.2.3 (Lipschitz estimates). *Let* $X : [0, T] \times \mathbb{R}^d \to \mathbb{R}^d$ *be a map. Then, for every* $\epsilon > 0$ *and every* $R > 0$, *we can find a set* $K \subset B_R(0)$ *such that* $\mathscr{L}^d(B_R(0) \setminus K) \le \epsilon$ *and for any* $0 \le t \le T$ *we have*

$$\mathrm{Lip}\, (X(t, \cdot)|_K) \le \exp \frac{c_d A_p(R, X)}{\epsilon^{1/p}} .$$

Proof. Fix $\epsilon > 0$ and $R > 0$. We can suppose that the quantity $A_p(R, X)$ is finite, otherwise the thesis is trivial; under this assumption, thanks to (A.7) we obtain a constant

$$M = M(\epsilon, p, A_p(R, X)) = \frac{A_p(R, X)}{\epsilon^{1/p}}$$

and a set $K \subset B_R(0)$ with $\mathscr{L}^d(B_R(0) \setminus K) \leq \epsilon$ and

$$\sup_{0 \leq t \leq T} \sup_{0 < r < 2R} \fint_{B_r(x)} \log\left(\frac{|X(t, x) - X(t, y)|}{r} + 1\right) dy \leq M \quad \forall x \in K.$$

This clearly means that

$$\fint_{B_r(x)} \log\left(\frac{|X(t, x) - X(t, y)|}{r} + 1\right) dy \leq M$$

for all $x \in K$, $t \in [0, T]$ and $r \in \,]0, 2R[$.

Now fix $x, y \in K$. Clearly $|x - y| < 2R$. Set $r = |x - y|$ and compute

$$\log\left(\frac{|X(t, x) - X(t, y)|}{r} + 1\right)$$

$$= \fint_{B_r(x) \cap B_r(y)} \log\left(\frac{|X(t, x) - X(t, y)|}{r} + 1\right) dz$$

$$\leq \fint_{B_r(x) \cap B_r(y)} \log\left(\frac{|X(t, x) - X(t, z)|}{r} + 1\right)$$

$$+ \log\left(\frac{|X(t, y) - X(t, z)|}{r} + 1\right) dz$$

$$\leq c_d \fint_{B_r(x)} \log\left(\frac{|X(t, x) - X(t, z)|}{r} + 1\right) dz$$

$$+ c_d \fint_{B_r(y)} \log\left(\frac{|X(t, y) - X(t, z)|}{r} + 1\right) dz$$

$$\leq c_d M = \frac{c_d A_p(R, X)}{\epsilon^{1/p}}.$$

This implies that

$$|X(t, x) - X(t, y)| \leq \exp\left(\frac{c_d A_p(R, X)}{\epsilon^{1/p}}\right) |x - y| \quad \text{for every } x, y \in K.$$

Therefore

$$\mathrm{Lip}(X(t, \cdot)|_K) \leq \exp \frac{c_d A_p(R, X)}{\epsilon^{1/p}}. \qquad \square$$

Remark 7.2.4. The quantitative Lipschitz estimate also holds under the assumption $b \in L^1([0, T]; W^{1,1}(\mathbb{R}^d; \mathbb{R}^d)) \cap L^\infty([0, T] \times \mathbb{R}^d; \mathbb{R}^d)$ and $M_\lambda Db \in L^1([0, T]; L^1(\mathbb{R}^d))$ for every $\lambda > 0$. To see this we define

$$\Phi(x) = \int_0^T M_{\tilde{R}} Db(t, X(t, x)) \, dt$$

and we go back to (7.5), which can be rewritten as

$$\sup_{0 \le t \le T} \sup_{0 < r < 2R} Q(t, x, r) \le c + c_d \Phi(x) + c_d M_{2R} \Phi(x).$$

For $\epsilon < 1/(4c)$ we can estimate

$$\mathscr{L}^d \left(\left\{ x \in B_R(0) \ : \ c + c_d \Phi(x) + c_d M_{2R} \Phi(x) > \frac{1}{\epsilon} \right\} \right)$$

$$\le \mathscr{L}^d \left(\left\{ x \in B_R(0) \ : \ c_d \Phi(x) > \frac{1}{4\epsilon} \right\} \right)$$

$$+ \mathscr{L}^d \left(\left\{ x \in B_R(0) \ : \ c_d M_{2R} \Phi(x) > \frac{1}{2\epsilon} \right\} \right)$$

$$\le \epsilon c_d \int_{B_R(0)} \Phi(x) \, dx + \epsilon c_d \int_{B_{3R}(0)} \Phi(x) \, dx$$

$$\le \epsilon c_d \int_0^T \int_{B_{3R}(0)} M_{\tilde{R}} Db(t, X(t, x)) \, dx \, dt$$

$$\le \epsilon c_d L \int_0^T \int_{B_{3R+T\|b\|_\infty}(0)} M_{\tilde{R}} Db(t, x) \, dx \, dt,$$

where in the third line we applied the Chebyshev inequality and the weak estimate (A.6) and in the last line Definition 7.1.1(ii). This means that it is possible to find a set $K \subset B_R(0)$ with $\mathscr{L}^d(B_R(0) \setminus K) \le \epsilon$ such that

$$\fint_{B_r(x)} \log \left(\frac{|X(t, x) - X(t, y)|}{r} + 1 \right) dy$$

$$\le \frac{c_d L}{\epsilon} \int_0^T \int_{B_{3R+T\|b\|_\infty}(0)} M_{\tilde{R}} Db(t, x) \, dx \, dt$$

for every $x \in K$, $t \in [0, T]$ and $r \in]0, 2R[$. Arguing as in the final part of the proof of Proposition 7.2.3 we obtain the Lipschitz estimate also in this case.

7.2.2. Existence, regularity and compactness

In this subsection we collect three direct corollaries of the estimates derived above, concerning approximate differentiability, existence and compactness of regular Lagrangian flows.

Corollary 7.2.5 (Approximate differentiability of the flow). *Let b be a bounded vector field belonging to $L^1([0, T]; W^{1,p}(\mathbb{R}^d; \mathbb{R}^d))$ for some $p > 1$, or belonging to $L^1([0, T]; W^{1,1}(\mathbb{R}^d; \mathbb{R}^d))$ and satisfying $M_\lambda Db \in L^1([0, T]; L^1(\mathbb{R}^d))$ for every $\lambda > 0$, and let X be a regular Lagrangian flow associated to b. Then $X(t, \cdot)$ is approximately differentiable $\mathscr{L}^d$-a.e. in $\mathbb{R}^d$, for every $t \in [0, T]$.*

Proof. The proof is an immediate consequence of the Lusin type approximation of the flow with Lipschitz maps given in Proposition 7.2.3 and Remark 7.2.4 and of Theorem A.2.3. $\qquad\square$

We now prove an "abstract" compactness lemma, which will be used immediately after to show a result of compactness of the flow. We prefer to state this lemma separately and in a more general way since we will refer to it also during the proofs of the other compactness results.

Lemma 7.2.6. *Let $\Omega \subset \mathbb{R}^n$ be a bounded Borel set and let $\{f_h\}$ be a sequence of maps into $\mathbb{R}^m$. Suppose that for every $\delta > 0$ we can find a positive constant $M_\delta < \infty$ and, for every fixed h, a Borel set $B_{h,\delta} \subset \Omega$ with $\mathscr{L}^n(\Omega \setminus B_{h,\delta}) \leq \delta$ in such a way that*

$$\|f_h\|_{L^\infty(B_{h,\delta})} \leq M_\delta$$

and

$$\mathrm{Lip}\left(f_h|_{B_{h,\delta}}\right) \leq M_\delta .$$

Then the sequence $\{f_h\}$ is precompact in measure in Ω.

Proof. For every $j \in \mathbb{N}$ we find the value $M_{1/j}$ and the sets $B_{h,1/j}$ as in the assumption of the lemma, with $\delta = 1/j$. Now, arguing component by component, we can extend every map $f_h|_{B_{h,1/j}}$ to a map f_h^j defined on Ω in such a way that the equi-bounds are preserved, up to a dimensional constant: we have

$$\|f_h^j\|_{L^\infty(\Omega)} \leq M_{1/j} \quad \text{for every } h$$

and

$$\mathrm{Lip}\left(f_h^j\right) \leq c_n M_{1/j} \quad \text{for every } h.$$

Then we apply the Ascoli-Arzelà theorem (notice that by uniform continuity all the maps f_h^j can be extended to the compact set $\bar{\Omega}$) and using a diagonal procedure we find a subsequence (in h) such that for every j the sequence $\{f_h^j\}_h$ converges uniformly in Ω to a map f_∞^j.

Now we fix $\epsilon > 0$. We choose $j \geq 3/\epsilon$ and we find $N = N(j)$ such that

$$\int_\Omega |f_i^j - f_k^j|\, dx \leq \epsilon/3 \quad \text{for every } i, k > N.$$

Keeping j and $N(j)$ fixed we estimate, for $i, k > N$

$$\int_\Omega 1 \wedge |f_i - f_k|\, dx \leq \int_\Omega 1 \wedge |f_i - f_i^j|\, dx$$
$$+ \int_\Omega 1 \wedge |f_i^j - f_k^j|\, dx + \int_\Omega 1 \wedge |f_k^j - f_k|\, dx$$
$$\leq \mathscr{L}^n(\Omega \setminus B_{i,1/j}) + \int_\Omega |f_i^j - f_k^j|\, dx + \mathscr{L}^n(\Omega \setminus B_{k,1/j})$$
$$\leq \frac{1}{j} + \frac{\epsilon}{3} + \frac{1}{j} \leq \epsilon.$$

It follows that the given sequence has a subsequence which is Cauchy with respect to the convergence in measure in Ω. This implies the thesis.   □

We now go back to the compactness of the flow.

Corollary 7.2.7 (Compactness of the flow). *Let $\{b_h\}$ be a sequence of vector fields equi-bounded in $L^\infty([0, T] \times \mathbb{R}^d; \mathbb{R}^d)$ and in $L^1([0, T]; W^{1,p}(\mathbb{R}^d; \mathbb{R}^d))$ for some $p > 1$, or equi-bounded in $L^\infty([0, T] \times \mathbb{R}^d; \mathbb{R}^d)$ and in $L^1([0, T]; W^{1,1}(\mathbb{R}^d; \mathbb{R}^d))$ and such that $\{M_\lambda Db_h\}$ is equi-bounded in $L^1([0, T]; L^1(\mathbb{R}^d))$ for every $\lambda > 0$. For each h, let X_h be a regular Lagrangian flow associated to b_h and let L_h be the compressibility constant of X_h, as in Definition 7.1.1(ii). Suppose that the sequence $\{L_h\}$ is equi-bounded. Then the sequence $\{X_h\}$ is strongly precompact in $L^1_{\text{loc}}([0, T] \times \mathbb{R}^d)$.*

Proof. Fix $\delta > 0$ and $R > 0$. Since $\{b_h\}$ is equi-bounded in $L^\infty([0, T] \times \mathbb{R}^d)$, we deduce that $\{X_h\}$ is equi-bounded in $L^\infty([0, T] \times B_R(0))$: let $C_1(R)$ be an upper bound for these norms. Applying Proposition 7.2.3, for every h we find a Borel set $K_{h,\delta}$ such that $\mathscr{L}^d(B_R(0) \setminus K_{h,\delta}) \leq \delta$ and

$$\text{Lip}\left(X_h(t, \cdot)|_{K_{h,\delta}}\right) \leq \exp \frac{c_d A_p(R, X_h)}{\delta^{1/p}} \quad \text{for every } t \in [0, T].$$

Recall first Theorem 7.2.1 implies that $A_p(R, X_h)$ is equi-bounded with respect to h, because of the assumptions of the corollary. Moreover, using Definition 7.1.1(i) and thanks again to the equi-boundedness of $\{b_h\}$ in $L^\infty([0, T] \times \mathbb{R}^d)$, we deduce that there exists a constant $C_2^\delta(R)$ such that

$$\mathrm{Lip}\left(X_h|_{[0,T] \times K_{h,\delta}}\right) \le C_2^\delta(R).$$

If we now set $B_{h,\delta} = [0, T] \times K_{h,\delta}$ and $M_\delta = \max\{C_1(R), C_2^\delta(R)\}$, we are in the position to apply Lemma 7.2.6 with $\Omega = [0, T] \times B_R(0)$. Then the sequence $\{X_h\}$ is precompact in measure in $[0, T] \times B_R(0)$, and by equi-boundedness in L^∞ we deduce that it is also precompact in $L^1([0, T] \times B_R(0))$. Using a standard diagonal argument it is possible to conclude that $\{X_h\}$ is locally precompact in $L^1([0, T] \times \mathbb{R}^d)$. $\square$

Corollary 7.2.8 (Existence of the flow). *Let b be a bounded vector field belonging to $L^1([0, T]; W^{1,p}(\mathbb{R}^d; \mathbb{R}^d))$ for some $p > 1$, or belonging to $L^1([0, T]; W^{1,1}(\mathbb{R}^d; \mathbb{R}^d))$ and satisfying $M_\lambda Db \in L^1([0, T]; L^1(\mathbb{R}^d))$ for every $\lambda > 0$. Assume that $[\mathrm{div}\, b]^- \in L^1([0, T]; L^\infty(\mathbb{R}^d))$. Then there exists a regular Lagrangian flow associated to b.*

Proof. This is a simple consequence of the previous corollary. Choose a positive convolution kernel in $\mathbb{R}^d$ and regularize b by convolution. It is simple to check that the sequence of smooth vector fields $\{b_h\}$ we have constructed satisfies the equi-bounds of the previous corollary. Moreover, since every b_h is smooth, for every h there is a unique regular Lagrangian flow associated to b_h, with compressibility constant L_h given by

$$L_h = \exp\left(\int_0^T \|[\mathrm{div}\, b_h(t, \cdot)]^-\|_{L^\infty(\mathbb{R}^d)}\, dt\right).\qquad(7.10)$$

Thanks to the positivity of the chosen convolution kernel, the sequence $\{L_h\}$ is equi-bounded, then we can apply Corollary 7.2.7. It is then easy to check that every limit point of $\{X_h\}$ in $L^1_{\mathrm{loc}}([0, T] \times \mathbb{R}^d)$ is a regular Lagrangian flow associated to b. $\square$

Remark 7.2.9. An analogous existence result could be obtained removing the hypothesis on the divergence of b, and assuming that there is some approximation procedure such that we can regularize b with equi-bounds on the compressibility constants of the approximating flows. This remark also applies to Corollaries 7.3.7 and 7.4.4.

7.2.3. Stability estimates and uniqueness

In this subsection we show an estimate similar in spirit to that of Theorem 7.2.1, but comparing flows for different vector fields. A direct corollary of this estimate is the stability (and hence the uniqueness) of regular Lagrangian flows.

Theorem 7.2.10 (Stability of the flow). *Let b and $\tilde{b}$ be bounded vector fields belonging to $L^1([0, T]; W^{1,p}(\mathbb{R}^d; \mathbb{R}^d))$ for some $p > 1$. Let X and $\tilde{X}$ be regular Lagrangian flows associated to b and $\tilde{b}$ respectively and denote by L and $\tilde{L}$ the compressibility constants of the flows. Then, for every time $\tau \in [0, T]$, we have*

$$\|X(\tau, \cdot) - \tilde{X}(\tau, \cdot)\|_{L^1(B_r(0))} \leq C \left|\log\left(\|b - \tilde{b}\|_{L^1([0,\tau]\times B_R(0))}\right)\right|^{-1},$$

where $R = r + T\|b\|_\infty$ and the constant C only depends on τ, r, $\|b\|_\infty$, $\|\tilde{b}\|_\infty$, L, $\tilde{L}$, and $\|D_x b\|_{L^1(L^p)}$.

Proof. Set $\delta = \|b - \tilde{b}\|_{L^1([0,\tau]\times B_R(0))}$ and consider the function

$$g(t) = \int_{B_r(0)} \log\left(\frac{|X(t, x) - \tilde{X}(t, x)|}{\delta} + 1\right) dx.$$

Clearly $g(0) = 0$ and after some standard computations we get

$$g'(t) \leq \int_{B_r(0)} \left|\frac{dX(t, x)}{dt} - \frac{d\tilde{X}(t, x)}{dt}\right| \left(|X(t, x) - \tilde{X}(t, x)| + \delta\right)^{-1} dx$$

$$= \int_{B_r(0)} \frac{|b(t, X(t, x)) - \tilde{b}(t, \tilde{X}(t, x))|}{|X(t, x) - \tilde{X}(t, x)| + \delta} dx$$

$$\leq \frac{1}{\delta} \int_{B_r(0)} |b(t, \tilde{X}(t, x)) - \tilde{b}(t, \tilde{X}(t, x))| dx$$

$$+ \int_{B_r(0)} \frac{|b(t, X(t, x)) - b(t, \tilde{X}(t, x))|}{|X(t, x) - \tilde{X}(t, x)| + \delta} dx. \tag{7.11}$$

We set $\tilde{R} = 2r + T(\|b\|_\infty + \|\tilde{b}\|_\infty)$ and we apply Lemma A.4.3 to estimate the last integral as follows:

$$\int_{B_r(0)} \frac{|b(t, X(t, x)) - b(t, \tilde{X}(t, x))|}{|X(t, x) - \tilde{X}(t, x)| + \delta} dx \leq c_d \int_{B_r(0)} M_{\tilde{R}} Db(t, X(t, x))$$

$$+ M_{\tilde{R}} Db(t, \tilde{X}(t, x)) dx.$$

Inserting this estimate in (7.11), setting $\tilde{r} = r + T \max\{\|b\|_\infty, \|\tilde{b}\|_\infty\}$, changing variables in the integrals and using Lemma A.4.2 we get

$$g'(t) \leq \frac{\tilde{L}}{\delta} \int_{B_{r+T\|\tilde{b}\|_\infty}(0)} |b(t, y) - \tilde{b}(t, y)| \, dy$$

$$+ (\tilde{L} + L) \int_{B_{\tilde{r}}(0)} M_{\tilde{R}} Db(t, y) \, dy$$

$$\leq \frac{\tilde{L}}{\delta} \int_{B_{r+T\|\tilde{b}\|_\infty}(0)} |b(t, y) - \tilde{b}(t, y)| \, dy$$

$$+ c_d \tilde{r}^{n-n/p} (\tilde{L} + L) \|M_{\tilde{R}} Db(t, \cdot)\|_{L^p}$$

$$\leq \frac{\tilde{L}}{\delta} \int_{B_{r+T\|\tilde{b}\|_\infty}(0)} |b(t, y) - \tilde{b}(t, y)| \, dy$$

$$+ c_{d,p} \tilde{r}^{n-n/p} (\tilde{L} + L) \|Db(t, \cdot)\|_{L^p} .$$

For any $\tau \in [0, T]$, integrating the last inequality between 0 and τ we get

$$g(\tau) = \int_{B_r(0)} \log\left(\frac{|X(\tau, x) - \tilde{X}(\tau, x)|}{\delta} + 1\right) dx \leq C_1, \quad (7.12)$$

where the constant C_1 depends on $\tau, r, \|b\|_\infty, \|\tilde{b}\|_\infty, L, \tilde{L}$, and $\|D_x b\|_{L^1(L^p)}$.

Next we fix a second parameter $\eta > 0$ to be chosen later. Using the Chebyshev inequality we find a measurable set $K \subset B_r(0)$ such that $\mathscr{L}^d(B_r(0) \setminus K) \leq \eta$ and

$$\log\left(\frac{|X(\tau, x) - \tilde{X}(\tau, x)|}{\delta} + 1\right) \leq \frac{C_1}{\eta} \qquad \text{for } x \in K.$$

Therefore we can estimate

$$\int_{B_r(0)} |X(\tau, x) - \tilde{X}(\tau, x)| \, dx$$

$$\leq \eta \left(\|X(\tau, \cdot)\|_{L^\infty(B_r(0))} + \|\tilde{X}(\tau, \cdot)\|_{L^\infty(B_r(0))}\right)$$

$$+ \int_K |X(\tau, x) - \tilde{X}(\tau, x)| \, dx$$

$$\leq \eta C_2 + c_d r^d \delta \left(\exp(C_1/\eta)\right) \leq C_3 \left(\eta + \delta \exp(C_1/\eta)\right),$$

(7.13)

with C_1, C_2 and C_3 which depend only on T, r, $\|b\|_\infty$, $\|\tilde{b}\|_\infty$, L, $\tilde{L}$, and $\|D_x b\|_{L^1(L^p)}$. Without loss of generality we can assume $\delta < 1$. Setting

$$\eta = 2C_1 |\log \delta|^{-1} = 2C_1 (-\log \delta)^{-1},$$

we have $\exp(C_1/\eta) = \delta^{-1/2}$. Thus we conclude

$$\int_{B_r(0)} |X(\tau, x) - \tilde{X}(\tau, x)| \, dx \leq C_3 \left(2C_1 |\log \delta|^{-1} + \delta^{1/2}\right)$$

$$\leq C |\log \delta|^{-1}, \tag{7.14}$$

where C depends only on τ, r, $\|b\|_\infty$, $\|\tilde{b}\|_\infty$, L, $\tilde{L}$, and $\|D_x b\|_{L^1(L^p)}$. This completes the proof.   □

Corollary 7.2.11 (Uniqueness of the flow). *Let b be a bounded vector field belonging to $L^1([0, T]; W^{1,p}(\mathbb{R}^d; \mathbb{R}^d))$ for some $p > 1$. Then the regular Lagrangian flow associated to b, if it exists, is unique.*

Proof. It follows immediately from the stability proved in Theorem 7.2.10.   □

Remark 7.2.12 (Stability with weak convergence in time). Theorem 7.2.10 allows to show the stability when the convergence of the vector fields is just weak with respect to the time. This setting is in fact very natural in view of the applications to the theory of fluid mechanics (see [84, Theorem II.7] and [107, Theorem 2.5]). In particular, under suitable bounds on the sequence $\{b_h\}$, the following form of weak convergence with respect to the time is sufficient to get the thesis:

$$\int_0^T b_h(t,x)\eta(t)dt \longrightarrow \int_0^T b(t,x)\eta(t)dt \text{ in } L^1_{\text{loc}}(\mathbb{R}^d) \text{ for every } \eta \in C_c^\infty(0,T).$$

Indeed, fix a parameter $\epsilon > 0$ and regularize with respect to the spatial variable only using a standard convolution kernel ρ_ϵ. We can rewrite the difference $X_h(t, x) - X(t, x)$ as

$$X_h(t, x) - X(t, x) = \left(X_h(t, x) - X_h^\epsilon(t, x)\right) + \left(X_h^\epsilon(t, x) - X^\epsilon(t, x)\right)$$

$$+ \left(X^\epsilon(t, x) - X(t, x)\right),$$

where X^ϵ and X_h^ϵ are the flows relative to the regularized vector fields b^ϵ and b_h^ϵ respectively. Now, it is simple to check that

- The last term goes to zero with ϵ, by the classical stability theorem (the quantitative version is not needed at this point);

- The first term goes to zero with ϵ, uniformly with respect to h: this is due to the fact that the difference $b_h^\epsilon - b_h$ goes to zero in $L_{loc}^1([0, T] \times \mathbb{R}^d)$ uniformly with respect to h, if we assume a uniform control in $W^{1,p}$ on the vector fields $\{b_h\}$, hence we can apply Theorem 7.2.10, and we get the desired convergence;
- The second term goes to zero for $h \to \infty$ when ϵ is kept fixed, because we are dealing with flows relative to vector fields which are smooth with respect to the space variable, uniformly in time, and weak convergence with respect to the time is enough to get the stability.

In order to conclude, we fix an arbitrary $\delta > 0$ and we first find $\epsilon > 0$ such that the norm of the third term is smaller than δ and such that the norm of the first term is smaller than δ for every h. For this fixed ϵ, we find h such that the norm of the second term is smaller than δ. With this choice of h we have estimated the norm of $X_h(t, x) - X(t, x)$ with 3δ, hence we get the desired convergence.

Remark 7.2.13 (Another way to show compactness). If we apply Theorem 7.2.10 to the flows $X(t, x)$ and $\tilde{X}(t, x) = X(t, x + h) - h$ relative to the vector fields $b(t, x)$ and $\tilde{b}(t, x) = b(t, x + h)$, where $h \in \mathbb{R}^d$ is fixed, we get for every $\tau \in [0, T]$

$$\|X(\tau, \cdot) - X(\tau, \cdot + h) - h\|_{L^1(B_r(0))}$$
$$\leq C \left|\log\left(\|b(t, x) - b(t, x + h)\|_{L^1([0,\tau]\times B_R(0))}\right)\right|^{-1}$$
$$\leq \frac{C}{|\log(h)|}.$$

Hence we have a uniform control on the translations in the space, and we can deduce a compactness result applying the Riesz-Fréchet-Kolmogorov compactness criterion (Lemma 7.4.2).

7.3. Estimates for more general vector fields and corollaries

In this section we extend the previous results to more general vector fields, in particular we drop the boundedness condition on b. More precisely, we will consider vector fields $b : [0, T] \times \mathbb{R}^d \to \mathbb{R}^d$ satisfying the following regularity assumptions:

(R1) $b \in L^1([0, T]; W_{loc}^{1,p}(\mathbb{R}^d; \mathbb{R}^d))$ for some $p > 1$;

(R2) We can write
$$\frac{b(t, x)}{1 + |x|} = \tilde{b}_1(t, x) + \tilde{b}_2(t, x)$$

with $\tilde{b}_1(t, x) \in L^1([0, T]; L^1(\mathbb{R}^d; \mathbb{R}^d))$ and $\tilde{b}_2(t, x) \in L^1([0, T]; L^\infty(\mathbb{R}^d; \mathbb{R}^d))$.

Since we are now considering vector fields which are no more bounded, we have to take care of the fact that the flow will be no more locally bounded in $\mathbb{R}^d$. However, we can give an estimate of the measure of the set of the initial data such that the corresponding trajectories exit from a fixed ball at some time.

Definition 7.3.1 (Sublevels). Fix $\lambda > 0$ and let $X : [0, T] \times \mathbb{R}^d \to \mathbb{R}^d$ be a locally summable map. We set

$$G_\lambda = \left\{ x \in \mathbb{R}^d \; : \; |X(t, x)| \leq \lambda \quad \forall t \in [0, T] \right\}. \tag{7.15}$$

Proposition 7.3.2 (Uniform estimate of the superlevels). *Let b be a vector field satisfying assumption (R2) and let X be a regular Lagrangian flow associated to b, with compressibility constant L. Then we have*

$$\mathscr{L}^d(B_R(0) \setminus G_\lambda) \leq g(R, \lambda),$$

where the function g only depends on $\|\tilde{b}_1\|_{L^1(L^1)}$, $\|\tilde{b}_2\|_{L^1(L^\infty)}$ *and L; moreover we have that* $g(R, \lambda) \downarrow 0$ *for R fixed and* $\lambda \uparrow +\infty$.

Proof. Let ϕ_t be the density of $X(t, \cdot)_{\#}(\mathbf{1}_{B_R(0)}\mathscr{L}^d)$ with respect to $\mathscr{L}^d$ and notice that, by the definition of push-forward and by Definition 7.1.1(ii), we have $\|\phi_t\|_1 = \omega_d R^d$ and $\|\phi_t\|_\infty \leq L$. Thanks to Definition 7.1.1(i) we can compute

$$\int_{B_R(0)} \sup_{0 \leq t \leq T} \log \left(\frac{1 + |X(t, x)|}{1 + R} \right) dx \leq \int_{B_R(0)} \int_0^T \frac{\left|\frac{dX}{dt}(t, x)\right|}{1 + |X(t, x)|} \, dt dx$$

$$= \int_0^T \int_{B_R(0)} \frac{|b(t, X(t, x))|}{1 + |X(t, x)|} \, dx dt$$

$$\leq \int_0^T \int_{\mathbb{R}^d} \frac{|b(t, x)|}{1 + |x|} \phi_t \, dx dt.$$

Using the Hölder inequality, for every decomposition of $b(t, x)/(1 + |x|)$ as in assumption (R2) we get

$$\int_{B_R(0)} \sup_{0 \leq t \leq T} \log \left(\frac{1 + |X(t, x)|}{1 + R} \right) dx \leq L\|\tilde{b}_1\|_{L^1(L^1)} + \omega_d R^d \|\tilde{b}_2\|_{L^1(L^\infty)}.$$

From this estimate we easily obtain

$$\mathscr{L}^d(B_R(0) \setminus G_\lambda) \leq \left[\log \left(\frac{1 + \lambda}{1 + R} \right) \right]^{-1} \left(L\|\tilde{b}_1\|_{L^1(L^1)} + \omega_d R^d \|\tilde{b}_2\|_{L^1(L^\infty)} \right),$$

and the right hand side clearly has the properties of the function $g(R, \lambda)$ stated in the proposition. $\qquad \square$

7.3.1. Estimate of an integral quantity and Lipschitz estimates

We start with the definition of an integral quantity which is a generalization of the quantity $A_p(R, X)$ of Theorem 7.2.1. In this new setting we will need a third variable (the truncation parameter λ), hence we define $A_p(R, \lambda, X)$ to be

$$\left[\int_{B_R(0) \cap G_\lambda} \left(\sup_{0 \le t \le T} \sup_{0 < r < 2R} \fint_{B_r(x) \cap G_\lambda} \log\left(\frac{|X(t,x) - X(t,y)|}{r} + 1 \right) dy \right)^p dx \right]^{\frac{1}{p}},$$
(7.16)

where the set G_λ is the sublevel relative to the map X, defined as in Definition 7.3.1.

In the following proposition, we show a bound on the quantity $A_p(R, \lambda, X)$ which corresponds to the bound on $A_p(R, X)$ in Theorem 7.2.1.

Theorem 7.3.3. *Let b be a vector field satisfying assumptions* (R1) *and* (R2) *and let X be a regular Lagrangian flow associated to b, with compressibility constant L. Then we have*

$$A_p(R, \lambda, X) \le C\left(R, L, \|D_x h\|_{L^1([0,T], L^p(B_{3\lambda}(0)))} \right).$$

Proof. We start as in the proof of Theorem 7.2.1, obtaining the validity of inequality (7.3) for every $x \in G_\lambda$. Since $|X(t, x) - X(t, y)| \le 2\lambda$, applying Lemma A.4.3 we deduce

$$\frac{dQ}{dt}(t, x, r) \le c_d M_{2\lambda} Db(t, X(t, x)) + c_d \fint_{B_r(x) \cap G_\lambda} M_{2\lambda} Db(t, X(t, y)) dy.$$

Then, arguing exactly as in the proof of Theorem 7.2.1, we get the estimate

$$A_p(R, \lambda, X)$$

$$\le c_{p,R} + c_d \left\| \int_0^T M_{2\lambda} Db(t, X(t, x)) dt \right\|_{L^p(B_R(0) \cap G_\lambda)}$$
(7.17)

$$+ c_d \left\| \int_0^T \sup_{0 < r < 2R} \fint_{B_r(x) \cap G_\lambda} M_{2\lambda} Db(t, X(t, y)) \, dy \, dt \right\|_{L^p(B_R(0) \cap G_\lambda)}$$
(7.18)

Recalling Definition 7.1.1(ii) and Lemma A.4.2, the integral in (7.17) can be estimated with

$$c_d L^{1/p} \int_0^T \|M_{2\lambda} Db(t,x)\|_{L^p(B_\lambda(0))} \, dt$$

$$\leq c_{d,p} L^{1/p} \int_0^T \|Db(t,x)\|_{L^p(B_{3\lambda}(0))} \, dt \,.$$

The integral in (7.18) can be estimated in a similar way with

$$c_d \int_0^T \left\| \sup_{0<r<2R} \fint_{B_r(x)\cap G_\lambda} [(M_{2\lambda} Db) \circ (t, X(t,\cdot))] \, (y) \, dy \right\|_{L^p(B_R(0)\cap G_\lambda)} dt$$

$$\leq c_d \int_0^T \left\| \sup_{0<r<2R} \fint_{B_r(x)} [(M_{2\lambda} Db) \circ (t, X(t,\cdot))] \, (y) \mathbf{1}_{G_\lambda}(y) dy \right\|_{L^p(B_R(0)\cap G_\lambda)} dt$$

$$= c_d \int_0^T \left\| M_{2R} \left[(M_{2\lambda} Db) \circ (t, X(t,\cdot)) \mathbf{1}_{G_\lambda}(\cdot) \right] (x) \right\|_{L^p(B_R(0)\cap G_\lambda)} dt$$

$$\leq c_{d,p} \int_0^T \left\| \left[(M_{2\lambda} Db) \circ (t, X(t,\cdot)) \mathbf{1}_{G_\lambda}(\cdot) \right] (x) \right\|_{L^p(B_{3R}(0))} dt$$

$$= c_{d,p} \int_0^T \left\| (M_{2\lambda} Db) \circ (t, X(t,x)) \right\|_{L^p(B_{3R}(0)\cap G_\lambda)} dt$$

$$\leq c_{d,p} L^{1/p} \int_0^T \|M_{2\lambda} Db(t,x)\|_{L^p(B_\lambda(0))} dt$$

$$\leq c_{d,p} L^{1/p} \int_0^T \|Db(t,x)\|_{L^p(B_{3\lambda}(0))} dt \,.$$

Then we obtain the desired estimate for $A_p(R, \lambda, X)$.   □

Proposition 7.3.4 (Lipschitz estimates). *Let X and b be as in Theorem 7.3.3. Then, for every $\epsilon > 0$ and every $R > 0$, we can find $\lambda > 0$ and a set $K \subset B_R(0)$ such that $\mathscr{L}^d(B_R(0) \setminus K) \leq \epsilon$ and for any $0 \leq t \leq T$ we have*

$$\mathrm{Lip}\,(X(t,\cdot)|_K) \leq \exp \frac{c_d A_p(R, \lambda, X)}{\epsilon^{1/p}} \,.$$

Proof. The proof is exactly the proof of Proposition 7.2.3, with some minor modifications due to the necessity of a truncation on the sublevels of the flow. This can be done as follows. For $\epsilon > 0$ and $R > 0$ fixed, we apply Proposition 7.3.2 to get a λ large enough such that $\mathscr{L}^d(B_R(0) \setminus G_\lambda) \leq \epsilon/2$. Next, using equation (A.7) and the finiteness of $A_p(R, \lambda, X)$, we obtain a constant

$$M = M(\epsilon, p, A_p(R, \lambda, X)) = \frac{A_p(R, \lambda, X)}{(\epsilon/2)^{1/p}}$$

and a set $K \subset B_R(0) \cap G_\lambda$ with $\mathscr{L}^d\big((B_R(0) \cap G_\lambda) \setminus K\big) \leq \epsilon/2$ and

$$\sup_{0 \leq t \leq T} \sup_{0 < r < 2R} \fint_{B_r(x) \cap G_\lambda} \log\left(\frac{|X(t, x) - X(t, y)|}{r} + 1\right) dy \leq M \quad \forall x \in K.$$

Hence the set K satisfies $\mathscr{L}^d(B_R(0) \setminus K) \leq \epsilon$ and

$$\fint_{B_i(x) \cap G_\lambda} \log\left(\frac{|X(t, x) - X(t, y)|}{r} + 1\right) dy \leq M$$

$$\forall x \in K, \forall t \in [0, T], \forall r \in]0, 2R[.$$

The proof can be concluded as the proof of Proposition 7.2.3, where now the integrals are performed on the sublevels G_λ. $\quad\square$

7.3.2. Existence, regularity and compactness

Corollary 7.3.5 (Approximate differentiability of the flow). *Let b be a vector field satisfying assumptions* (R1) *and* (R2) *and let X be a regular Lagrangian flow associated to b. Then $X(t, \cdot)$ is approximately differentiable $\mathscr{L}^d$-a.e. in $\mathbb{R}^d$, for every $t \in [0, T]$.*

Proof. The proof is an immediate consequence of the Lusin type approximation of the flow with Lipschitz maps given in Proposition 7.3.4 and of Theorem A.2.3. $\quad\square$

Corollary 7.3.6 (Compactness of the flow). *Let $\{b_h\}$ be a sequence of vector fields satisfying assumptions* (R1) *and* (R2). *For every h, let X_h be a regular Lagrangian flow associated to b_h and let L_h be the compressibility constant associated to X_h, as in Definition 7.1.1(ii). Suppose that for every $R > 0$ the uniform estimate*

$$\|D_x b_h\|_{L^1([0,T]; L^p(B_R(0)))} + \|\tilde{b}_{h,1}\|_{L^1(L^1)} + \|\tilde{b}_{h,2}\|_{L^1(L^\infty)} + L_h \leq C(R) < \infty \tag{7.19}$$

is satisfied, for some decomposition $b_h/(1 + |x|) = \tilde{b}_{h,1} + \tilde{b}_{h,2}$ as in assumption (R2). *Then the sequence $\{X_h\}$ is locally precompact in measure in $[0, T] \times \mathbb{R}^d$.*

Proof. The proof is essentially identical to the proof of Corollary 7.2.7. Fix $R > 0$ and $\delta > 0$. Applying Proposition 7.3.2 and thanks to the uniform bound given by (7.19), we first find $\lambda > 0$ big enough such that

$$\mathcal{L}^d\big(B_R(0) \setminus G_\lambda^h\big) \le \delta/3\,,$$

with G_λ^h as in Definition 7.3.1. Thanks again to (7.19), we can apply Theorem 7.3.3 to deduce that the quantities $A_p(R, \lambda, X_h)$ are uniformly bounded with respect to h. Now we apply Proposition 7.3.4 with $\epsilon = \delta/3$ to find, for every h, a measurable set $K_h \subset B_R(0) \cap G_\lambda^h$ such that

$$\mathcal{L}^d\big((B_R(0) \cap G_\lambda^h) \setminus K_h\big) \le \delta/3$$

and

$$\text{Lip}\left(X_h(t, \cdot)|_{K_h}\right) \quad \text{is uniformly bounded w.r.t. } h.$$

Now we are going to show a similar Lipschitz estimate with respect to the time. Since the maps

$$[0, T] \times K_h \ni (t, x) \mapsto b_h(t, X_h(t, x))$$

are uniformly bounded in $L^1([0, T] \times K_h)$ (this is easily deduced recalling assumption (R2), the bound (7.19) and the fact that $K_h \subset B_R(0)$), for every h, applying the Chebyshev inequality, we can find a measurable set $H_h \subset [0, T] \times K_h$ such that

$$\mathcal{L}^d\big(([0, T] \times K_h) \setminus H_h\big) \le \delta/3$$

and

$$\|b_h(t, X_h(t, x))\|_{L^\infty(H_h)} \le C/\delta\,,$$

where the constant C only depends on the constant $C(R)$ given by (7.19). Then we deduce that

$$\left\|\frac{dX_h}{dt}(t, x)\right\|_{L^\infty(H_h)} \quad \text{is uniformly bounded w.r.t. } h.$$

Hence we have found, for every h, a measurable set $H_h \subset [0, T] \times B_R(0)$ such that

$$\mathcal{L}^d\big(([0, T] \times B_R(0)) \setminus H_h\big) \le \delta$$

and

$$\|X_h\|_{L^\infty(H_h)} + \text{Lip}_{t,x}\left(X_h|_{H_h}\right) \quad \text{uniformly bounded w.r.t. } h.$$

Then we apply Lemma 7.2.6 to obtain that the sequence $\{X_h\}$ is precompact in measure in $[0, T] \times B_R(0)$. A standard diagonal argument gives the local precompactness in measure of the sequence in the whole $[0, T] \times \mathbb{R}^d$. $\qquad\square$

Corollary 7.3.7 (Existence of the flow). *Let b be a vector field satisfying assumptions* (R1) *and* (R2) *and such that* $[\operatorname{div} b]^{-} \in L^{1}([0, T];$ $L^{\infty}(\mathbb{R}^{d}))$. *Then there exists a regular Lagrangian flow associated to b.*

Proof. It is sufficient to regularize b with a positive convolution kernel in $\mathbb{R}^{d}$ and apply Corollary 7.3.6. It is simple to check that the regularized vector fields satisfy the equi-bounds needed for the compactness result.
$\cdot\ \square$

7.3.3. Stability estimates and uniqueness

Theorem 7.3.8 (Stability estimate). *Let b and $\tilde{b}$ be vector fields satisfying assumptions (R1) and (R2). Let X and $\tilde{X}$ be regular Lagrangian flows associated to b and $\tilde{b}$ respectively and denote by L and $\tilde{L}$ the compressibility constants of the flows. Then for every $\lambda > 1$ and every $\tau \in [0, T]$ the following estimate holds*

$$\int_{B_r(0)} 1 \wedge |X(\tau, x) - \tilde{X}(\tau, x)| \, dx \ \leq\ \frac{C}{\log(\lambda)} + C_{\lambda} \|b - \tilde{b}\|_{L^1([0,\tau] \times B_{\lambda}(0))} ,$$

$$(7.20)$$

where the constant C only depends on L, $\tilde{L}$ and on the $L^1(L^1) + L^1(L^{\infty})$ norm of some decomposition of b and $\tilde{b}$ as in assumption (R2), while the constant C_{λ} depends on λ, r, L, $\tilde{L}$ and $\|Db\|_{L^1([0,\tau]; L^p(B_{3\lambda}(0)))}$.

Proof. For any given $\lambda > 1$ fixed define the sets G_{λ} and $\tilde{G}_{\lambda}$, relatively to X and $\tilde{X}$, as in (7.15). Set

$$\delta = \delta(\lambda) = \|b - \tilde{b}\|_{L^1([0,\tau] \times B_{\lambda}(0))} .$$

Define

$$g(t) = \int_{B_r(0) \cap G_{\lambda} \cap \tilde{G}_{\lambda}} \log\left(\frac{|X(t, x) - \tilde{X}(t, x)|}{\delta} + 1\right) dx .$$

Clearly we have $g(0) = 0$ and we can estimate

$$g'(t) \leq \int_{B_r(0) \cap G_{\lambda} \cap \tilde{G}_{\lambda}} \frac{|b(t, X(t, x)) - \tilde{b}(t, \tilde{X}(t, x))|}{|X(t, x) - \tilde{X}(t, x)| + \delta} dx$$

$$\leq \int_{B_r(0) \cap G_{\lambda} \cap \tilde{G}_{\lambda}} \frac{|b(t, \tilde{X}(t, x)) - \tilde{b}(t, \tilde{X}(t, x))|}{|X(t, x) - \tilde{X}(t, x)| + \delta}$$

$$+ \frac{|b(t, X(t, x)) - b(t, \tilde{X}(t, x))|}{|X(t, x) - \tilde{X}(t, x)| + \delta} dx$$

$$\leq \int_{B_r(0) \cap G_\lambda \cap \tilde{G}_\lambda} \frac{1}{\delta} |b(t, \tilde{X}(t, x)) - \tilde{b}(t, \tilde{X}(t, x))|$$

$$+ \frac{|b(t, X(t, x)) - b(t, \tilde{X}(t, x))|}{|X(t, x) - \tilde{X}(t, x)|} dx$$

$$\leq \frac{1}{\delta} \int_{B_r(0) \cap G_\lambda \cap \tilde{G}_\lambda} |b(t, \tilde{X}(t, x)) - \tilde{b}(t, \tilde{X}(t, x))| dx$$

$$+ c_d \int_{B_r(0) \cap G_\lambda \cap \tilde{G}_\lambda} \left(M_{2\lambda} Db(t, X(t, x)) + M_{2\lambda} Db(t, \tilde{X}(t, x)) \right) dx$$

$$\leq \frac{\tilde{L}}{\delta} \int_{B_\lambda(0)} |b(t, x) - \tilde{b}(t, x)| dx$$

$$+ c_d (L + \tilde{L}) \int_{B_\lambda(0)} M_{2\lambda} Db(t, x) dx$$

$$\leq \frac{\tilde{L}}{\delta} \int_{B_\lambda(0)} |b(t, x) - \tilde{b}(t, x)| dx$$

$$+ c_{d,p} (L + \tilde{L}) \lambda^{n-n/p} \|Db(t, \cdot)\|_{L^p(B_{3\lambda}(0))} .$$

Integrating with respect to t between 0 and τ we obtain

$$g(\tau) = \int_{B_r(0) \cap G_\lambda \cap \tilde{G}_\lambda} \log \left(\frac{|X(\tau, x) - \tilde{X}(\tau, x)|}{\delta} + 1 \right) dx$$

$$\leq \tilde{L} + c_{d,p} (L + \tilde{L}) \lambda^{n-n/p} \|Db\|_{L^1([0,\tau]; L^p(B_{3\lambda}(0)))}$$

$$= C_\lambda ,$$

where the constant C_λ depends on λ but also on the other parameters relative to b and $\tilde{b}$. Now fix a value $\eta > 0$ which will be specified later. We can find a measurable set $K \subset B_r(0) \cap G_\lambda \cap \tilde{G}_\lambda$ such that $\mathscr{L}^d \left((B_r(0) \cap G_\lambda \cap \tilde{G}_\lambda) \setminus K \right) < \eta$ and

$$\log \left(\frac{|X(\tau, x) - \tilde{X}(\tau, x)|}{\delta} + 1 \right) \leq \frac{C_\lambda}{\eta} \qquad \forall x \in K .$$

Then we deduce that

$$\int_{B_r(0)} 1 \wedge |X(\tau, x) - \tilde{X}(\tau, x)| \, dx$$

$$\leq \mathscr{L}^d \big(B_r(0) \setminus (G_\lambda \cap \tilde{G}_\lambda) \big)$$

$$+ \mathscr{L}^d \big((B_r(0) \cap G_\lambda \cap \tilde{G}_\lambda) \setminus K \big) + \int_K |X(t, x) - \tilde{X}(t, x)| \, dx$$

$$\leq \frac{C}{\log(\lambda)} + \eta + C\delta \exp(C_\lambda / \eta)$$

$$\leq \frac{C}{\log(\lambda)} + C_\lambda \| b - \tilde{b} \|_{L^1([0,\tau] \times B_\lambda(0))} ,$$

choosing $\eta = 1/\log(\lambda)$ in the last line. □

Corollary 7.3.9 (Stability of the flow). *Let $\{b_h\}$ be a sequence of vector fields satisfying assumptions (R1) and (R2), converging in $L^1_{loc}([0, T] \times \mathbb{R}^d)$ to a vector field b which satisfies assumptions (R1) and (R2). Denote by X and X_h the regular Lagrangian flows associated to b and b_h respectively, and denote by L and L_h the compressibility constants of the flows. Suppose that, for some decomposition $b_h/(1 + |x|) = \tilde{b}_{h,1} + \tilde{b}_{h,2}$ as in assumption (R2), we have*

$$\| \tilde{b}_{h,1} \|_{L^1(L^1)} + \| \tilde{b}_{h,2} \|_{L^1(L^\infty)} \qquad \text{equi-bounded in } h$$

and that the sequence $\{L_h\}$ is equi-bounded. Then the sequence $\{X_h\}$ converges to X locally in measure in $[0, T] \times \mathbb{R}^d$.

Proof. Notice that, under the hypotheses of this corollary, the constants $C^{h,\tau}$ and $C^{h,\tau}_\lambda$ in (7.20) can be chosen uniformly with respect to $\tau \in [0, T]$ and $h \in \mathbb{N}$. Hence we find universal constant C and C_λ, depending only on the assumed equi-bounds, such that

$$\int_{B_r(0)} 1 \wedge |X(\tau, x) - X_h(\tau, x)| \, dx$$

$$\leq \frac{C^{h,\tau}}{\log(\lambda)} + C^{h,\tau}_\lambda \| b - b_h \|_{L^1([0,\tau] \times B_\lambda(0))} \qquad (7.21)$$

$$\leq \frac{C}{\log(\lambda)} + C_\lambda \| b - b_h \|_{L^1([0,T] \times B_\lambda(0))} .$$

Now fix $\epsilon > 0$. We first choose λ big enough such that

$$\frac{C}{\log(\lambda)} \leq \frac{\epsilon}{2} ,$$

where C is the first constant in (7.21). Since now λ is fixed, we find N such that for every $h \geq N$ we have

$$\|b - b_h\|_{L^1([0,T] \times B_\lambda(0))} \leq \frac{\epsilon}{2C_\lambda},$$

thanks to the convergence of the sequence $\{b_h\}$ to b in $L^1_{\text{loc}}([0, T] \times \mathbb{R}^d)$. Notice that N depends on λ and on the equi-bounds, but in turn λ only depends on ϵ and on the equi-bounds. Hence we get

$$\int_{B_r(0)} 1 \wedge |X(\tau, x) - X_h(\tau, x)| \, dx \leq \epsilon \quad \text{for every } h \geq N = N(\epsilon).$$

This means that $\{X_h(\tau, \cdot)\}$ converges to $X(\tau, \cdot)$ locally in measure in $\mathbb{R}^d$, uniformly with respect to $\tau \in [0, T]$. In particular we get the thesis. $\square$

Corollary 7.3.10 (Uniqueness of the flow). *Let b be a vector field satisfying assumptions* (R1) *and* (R2). *Then the regular Lagrangian flow associated to b, if it exists, is unique.*

Proof. It follows immediately from Corollary 7.3.9. $\square$

7.4. A direct proof of compactness

In this section we propose an alternative proof of the compactness result of Theorem 7.2.7, which works under an assumption of summability of the maximal function of Db. The strategy of this proof is slightly different from the previous one: we are not going to use the Lipschitz estimates of Proposition 7.2.3 and Remark 7.2.4, but instead we prove an estimate of an integral quantity which turns out to be sufficient to get compactness, via the Riesz-Fréchet-Kolmogorov compactness criterion.

We will assume the following regularity assumption on the vector field:

(R3) For every $\lambda > 0$ we have $M_\lambda Db \in L^1([0, T]; L^1_{\text{loc}}(\mathbb{R}^d))$.

Notice that, by Lemma A.4.2, this assumption is equivalent to the condition

$$\int_0^T \int_{B_\rho(0)} |D_x b(t, x)| \log (2 + |D_x b(t, x)|) \, dx dt < \infty \quad \text{for every } \rho > 0.$$

This means that $D_x b \in L^1([0, T]; L \log L_{\text{loc}}(\mathbb{R}^d))$, *i.e.* a slightly stronger bound than $D_x b \in L^1([0, T], L^1_{\text{loc}}(\mathbb{R}^d))$.

We define a new integral quantity, which corresponds to the one defined in Theorem 7.2.1 for $p = 1$, but without the supremum with respect to r. For $R > 0$ and $0 < r < R/2$ fixed we set

$$a(r, R, X) = \int_{B_R(0)} \sup_{0 \leq t \leq T} \fint_{B_r(x)} \log \left(\frac{|X(t, x) - X(t, y)|}{r} + 1 \right) dy dx.$$

We first give a quantitative estimate for the quantity $a(r, R, X)$, similar to the one for $A_p(R, X)$.

Theorem 7.4.1. *Let b be a bounded vector field satisfying assumption* (R3) *and let X be a regular Lagrangian flow associated to b, with compressibility constant L. Then we have*

$$a(r, R, X) \le C \left(R, L, \| M_{\tilde{R}} D_x b \|_{L^1([0,T]; L^1(B_{\tilde{R}}(0)))} \right),$$

where $\tilde{R} = 3R/2 + 2T\|b\|_\infty$.

Proof. We start as in the proof of Theorem 7.2.1, obtaining inequality (7.4) (but this time it is sufficient to set $\tilde{R} = 3R/2 + 2T\|b\|_\infty$). Integrating with respect to the time and then with respect to x over $B_R(0)$, we obtain

$$a(r, R, X) \le c_R + c_d \int_{B_R(0)} \int_0^T M_{\tilde{R}} Db(t, X(t, x)) \, dt \, dx$$

$$+ c_d \int_{B_R(0)} \int_0^T \fint_{B_r(x)} M_{\tilde{R}} Db(t, X(t, y)) \, dy \, dt \, dx.$$

As in the previous computations, the first integral can be estimated with

$$c_d L \left\| M_{\tilde{R}} Db \right\|_{L^1([0,T]; L^1(B_{R+T\|b\|_\infty}(0)))},$$

but this time we cannot bound the norm of the maximal function with the norm of the derivative. To estimate the last integral we compute

$$c_d \int_{B_R(0)} \int_0^T \fint_{B_r(x)} M_{\tilde{R}} Db(t, X(t, y)) \, dy \, dt \, dx$$

$$= c_d \int_{B_R(0)} \int_0^T \fint_{B_r(0)} M_{\tilde{R}} Db(t, X(t, x + z)) \, dz \, dt \, dx$$

$$\le c_d \fint_{B_r(0)} \int_0^T \int_{B_R(0)} M_{\tilde{R}} Db(t, X(t, x + z)) \, dx \, dt \, dz$$

$$\le c_d \fint_{B_r(0)} \int_0^T L \int_{B_{3R/2 + T\|b\|_\infty}(0)} M_{\tilde{R}} Db(t, w) \, dw \, dt \, dx$$

$$= c_d L \| M_{\tilde{R}} Db \|_{L^1([0,T]; L^1(B_{3R/2 + T\|b\|_\infty}(0)))}.$$

Hence the thesis follows, by definition of $\tilde{R}$.   □

Next we show in Corollary 7.4.3 how this estimate implies compactness for the flow. We will need the following well-known criterion for strong compactness in L^p.

Lemma 7.4.2 (Riesz-Fréchet-Kolmogorov compactness criterion).
Let $\mathscr{F}$ be a bounded subset of $L^p(\mathbb{R}^N)$ for some $1 \le p < \infty$. Suppose that

$$\lim_{|h| \to 0} \|f(\cdot - h) - f\|_p = 0 \qquad \text{uniformly in } f \in \mathscr{F}.$$

Then $\mathscr{F}$ is relatively compact in $L^p_{\text{loc}}(\mathbb{R}^N)$.

Corollary 7.4.3 (Compactness of the flow). *Let $\{b_h\}$ be a sequence of vector fields equi-bounded in $L^\infty([0, T] \times \mathbb{R}^d; \mathbb{R}^d)$ and suppose that the sequence $\{M_\lambda Db_h\}$ is equi-bounded in $L^1([0, T]; L^1_{\text{loc}}(\mathbb{R}^d))$ for every $\lambda > 0$. For each h, let X_h be a regular Lagrangian flow associated to b_h and let L_h be the compressibility constant associated to X_h, as in Definition 7.1.1(ii). Suppose that the sequence $\{L_h\}$ is equi-bounded. Then the sequence $\{X_h\}$ is strongly precompact in $L^1_{\text{loc}}([0, T] \times \mathbb{R}^d).$*

Proof. We apply Theorem 7.4.1 to obtain that, under the assumptions of the corollary, the quantities $a(r, R, X_h)$ are uniformly bounded with respect to h. Now observe that, for $0 \le z \le \tilde{R}$ (with $\tilde{R} = 3R/2 + 2T \|b\|_\infty$ as in Theorem 7.4.1), thanks to the concavity of the logarithm we have

$$\log\left(\frac{z}{r} + 1\right) \ge \frac{\log\left(\frac{\tilde{R}}{r} + 1\right)}{\tilde{R}} z \, .$$

Since $|X_h(t, x) - X_h(t, y)| \le \tilde{R}$ this implies that

$$\int_{B_R(0)} \sup_{0 \le t \le T} \fint_{B_r(x)} |X_h(t, x) - X_d(t, y)| \, dy dx$$

$$\le \frac{\tilde{R}}{\log\left(\frac{\tilde{R}}{r} + 1\right)} C\left(R, L_h, \|M_{\tilde{R}} Db_h\|_{L^1([0,T];L^1(B_{\tilde{R}}(0)))}\right) \le g(r),$$

where the function $g(r)$ does not depend on h and satisfies $g(r) \downarrow 0$ for $r \downarrow 0$. Changing the integration order this implies

$$\fint_{B_r(0)} \int_{B_R(0)} |X_h(t, x) - X_h(t, x + z)| \, dx dz \le g(r) \, ,$$

uniformly with respect to t and h.

Now notice the following elementary fact. There exists a dimensional constant $\alpha_d > 0$ with the following property: if $A \subset B_1(0)$ is a measurable set with $\mathscr{L}^d(B_1(0) \setminus A) \le \alpha_d$, then $A + A \supset B_{1/2}(0)$. Indeed, if the thesis were false, we could find $x \in B_{1/2}(0)$ such that $x \notin A + A$. This would imply in particular that $x \notin (A \cap B_{1/2}(0)) + (A \cap B_{1/2}(0))$, so that

$$\left[x - \left(A \cap B_{1/2}(0) \right) \right] \cap \left[A \cap B_{1/2}(0) \right] = \emptyset. \qquad (7.22)$$

Now notice that there exists a dimensional constant γ_d such that

$$\mathcal{L}^d \left(B_{1/2}(0) \cap (x - B_{1/2}(0)) \right) \geq \gamma_d \,,$$

since we are supposing $x \in B_{1/2}(0)$. But since $\mathcal{L}^d (B_1(0) \setminus A) \leq \alpha_d$, we also have

$$\mathcal{L}^d \left(B_{1/2}(0) \setminus \left(A \cap B_{1/2}(0) \right) \right) \leq \alpha_d$$

and

$$\mathcal{L}^d \left((x - B_{1/2}(0)) \setminus (x - (A \cap B_{1/2}(0))) \right) = \mathcal{L}^d \left(B_{1/2}(0) \setminus (A \cap B_{1/2}(0)) \right) \leq \alpha_d.$$

But this is clearly in contradiction with (7.22) if we choose $\alpha_d < \gamma_d / 2$.

Then fix α_d as above and apply the Chebyshev inequality for every h to obtain, for every $0 < r < R/2$, a measurable set $K_{r,h} \subset B_r(0)$ with $\mathcal{L}^d \left(B_r(0) \setminus K_{r,h} \right) \leq \alpha_d \mathcal{L}^d (B_r(0))$ and

$$\int_{B_R(0)} |X_h(t, x + z) - X_h(t, x)| \, dx \leq \frac{g(r)}{\alpha_d} \qquad \text{for every } z \in K_{r,h}.$$

For such a set $K_{r,h}$, thanks to the previous remark, we have that $K_{r,h} + K_{r,h} \supset B_{r/2}(0)$. Now let $v \in B_{r/2}(0)$ be arbitrary. For every h we can write $v = z_{1,h} + z_{2,h}$ with $z_{1,h}, z_{2,h} \in K_{r,h}$. We can estimate the increment in the spatial directions as follows:

$$\int_{B_{R/2}(0)} |X_h(t, x + v) - X_h(t, x)| \, dx$$

$$= \int_{B_{R/2}(0)} |X_h(t, x + z_{1,h} + z_{2,h}) - X_h(t, x)| \, dx$$

$$\leq \int_{B_{R/2}(0)} |X_h(t, x + z_{1,h} + z_{2,h}) - X_h(t, x + z_{1,h})|$$

$$+ |X_h(t, x + z_{1,h}) - X_h(t, x)| \, dx$$

$$\leq \int_{B_R(0)} |X_h(t, y + z_{2,h}) - X_h(t, y)| \, dy$$

$$+ \int_{B_R(0)} |X_h(t, x + z_{1,h}) - X_h(t, x)| \, dx \leq \frac{2g(r)}{\alpha_d}.$$

Now notice that, by Definition 7.1.1(i), for $\mathscr{L}^d$-a.e. $x \in \mathbb{R}^d$ we have

$$\frac{dX_h}{dt}(t, x) = b_h(t, X_h(t, x)) \quad \text{for every } t \in [0, T].$$

Then we can estimate the increment in the time direction in the following way

$$|X_h(t + \tau, x) - X_h(t, x)| \leq \int_0^\tau \left| \frac{dX_h}{dt}(t + s, x) \right| ds$$

$$= \int_0^\tau |b_h(t + s, X_h(t + s, x))| ds \leq \tau \|b_h\|_\infty.$$

Combining these two informations, for $(t_0, t_1) \in [0, T]$, $R > 0$, $v \in B_{r/2}(0)$ and $\tau > 0$ sufficiently small we can estimate

$$\int_{t_0}^{t_1} \int_{B_{R/2}(0)} |X_h(t + \tau, x + v) - X_h(t, x)| \, dx dt$$

$$\leq \int_{t_0}^{t_1} \int_{B_{R/2}(0)} |X_h(t + \tau, x + v) - X_h(t + \tau, x)|$$

$$+ |X_h(t + \tau, x) - X_h(t, x)| \, dx dt$$

$$\leq T \frac{2g(r)}{\alpha_d} + \int_{t_0}^{t_1} \int_{B_{R/2}(0)} \tau \|b_h\|_\infty \, dx dt$$

$$\leq T \frac{2g(r)}{\alpha_d} + c_d T R^d \tau \|b_h\|_\infty.$$

The thesis follows applying the Riesz-Fréchet-Kolmogorov compactness criterion (see Lemma 7.4.2), recalling that $\{b_h\}$ is uniformly bounded in $L^\infty([0, T] \times \mathbb{R}^d)$. $\qquad \square$

Corollary 7.4.4 (Existence of the flow). *Let b be a bounded vector field satisfying assumption (R3) and such that $[\mathrm{div}\, b]^- \in L^1([0, T]; L^\infty(\mathbb{R}^d))$. Then there exists a regular Lagrangian flow associated to b.*

Proof. It is sufficient to regularize b with a positive convolution kernel in $\mathbb{R}^d$ and apply Corollary 7.4.3. It is simple to check that the regularized vector fields satisfy the equi-bounds needed for the compactness result, due to the convexity of the map $z \mapsto z \log(2 + z)$ for $z \geq 0$. $\qquad \square$

7.5. Lipexp$_p$–regularity for transport equations with $W^{1,p}$ coefficients

In this section we show that solutions to transport equations with Sobolev coefficients propagate a very mild regularity property of the initial data; compare the result of Theorem 7.5.3 with the examples of Section 5.4.

Definition 7.5.1 (The space Lipexp$_p$). We say that a function $f : E \in \mathbb{R}^d \to \mathbb{R}^k$ belongs to Lipexp$_p(E)$ if for every $\epsilon > 0$ there exists a measurable set $K \subset E$ such that

(i) $\mathscr{L}^d(E \setminus K) \leq \epsilon$;

(ii) $\mathrm{Lip}(f|_K) \leq \exp\left(C\epsilon^{-1/p}\right)$ for some constant $C < \infty$ independent on ϵ.

Moreover we denote by $|f|_{\mathrm{LE}_p(E)}$ the smallest constant C such that the conditions above hold.

Remark 7.5.2. Note that:

- Lipexp$_\infty$ is the space of functions which coincide with a Lipschitz function almost everywhere;
- $|f|_{\mathrm{LE}_p(E)}$ is not homogeneus, and then it is not a norm, and can be explicitely defined as

$$|f|_{\mathrm{LE}_p(E)} = \sup_{\epsilon>0} \left\{\epsilon^{1/p} \log \min \left\{\mathrm{Lip}(f|_K) \; : \; \mathscr{L}^d(E \setminus K) \leq \epsilon\right\}\right\} ;$$

- one can compare this definition with a similar result for Sobolev functions: if $f \in W^{1,p}(E; \mathbb{R}^k)$, then for every $\epsilon > 0$ there exists a set $K \subset E$ such that $\mathscr{L}^d(E \setminus K) \leq \epsilon$ and $\mathrm{Lip}(f|_K) \leq \|Df\|_{L^p(E)}\epsilon^{-1/p}$.

Theorem 7.5.3. *Let b be a vector field satisfying assumptions (R1) and (R2) and such that $\mathrm{div}\, b \in L^1([0, T]; L^\infty(\mathbb{R}^d))$. Let $\bar{u} \in L^\infty(\mathbb{R}^d)$ such that $\bar{u} \in \mathrm{Lipexp}_p(\Omega)$ for every $\Omega \Subset \mathbb{R}^d$. Let u be the solution of the Cauchy problem*

$$\begin{cases} \partial_t u(t, x) + b(t, x) \cdot \nabla u(t, x) = 0 \\ u(0, \cdot) = \bar{u} . \end{cases} \tag{7.23}$$

Then for every $\Omega \Subset \mathbb{R}^d$ we have that

$$\sup_{0 \leq t \leq T} |u(t, \cdot)|_{\mathrm{LE}_p(\Omega)} < \infty .$$

Remark 7.5.4. Recalling Remark 2.2.2 we observe that we can define $u(t, \cdot)$ for every $t \in [0, T]$ in such a way that $u \in C([0, T], L^1_{\mathrm{loc}}(\mathbb{R}^d) - w)$.

Proof of Theorem 7.5.3. Let X be the regular Lagrangian flow generated by b. Then:

(a) There exists a constant $C > 0$ such that $C^{-1}\mathscr{L}^d(\Omega) \leq \mathscr{L}^d\big(X(t, \Omega)\big) \leq C\mathscr{L}^d(\Omega)$ for every $t \in [0, T]$ and for every $\Omega \subset \mathbb{R}^d$; therefore, for every $t \in [0, T]$, we can define $\Psi(t, x)$ via the identity $X(t, \Psi(t, x)) = \Psi(t, X(t, x)) = x$ for $\mathscr{L}^d$-a.e. $x \in \mathbb{R}^d$;

(b) For every t we have $u(t, x) = \bar{u}(\Psi(t, x))$ for almost every x.

Note that if for every t we consider the regular Lagrangian flow $\Phi(t, \cdot, \cdot)$ of

$$\begin{cases} \dfrac{d\Phi}{d\tau}(t, \tau, x) = -b(t - \tau, \Phi(t, \tau, x)) \\[2mm] \Phi(t, 0, x) = x\,, \end{cases}$$

then $\Psi(t, x) = \Phi(t, t, x)$. Therefore, thanks to Proposition 7.3.4 we conclude that

$$\sup_{0 \leq t \leq T} |\Psi(t, \cdot)|_{\mathrm{LE}_p(\Omega)} \leq C_1(\Omega)$$

for every $\Omega \in \mathbb{R}^d$.

Let $t \in [0, T]$, $R > 0$ and $\epsilon > 0$ be given. Choose $K_1 \subset B_R(0)$ such that

- $\mathscr{L}^d\big(B_R(0) \setminus K_1\big) \leq \epsilon/3$,
- $\mathrm{Lip}(\Psi(t, \cdot)|_{K_1}) \leq \exp(|\Psi(t, \cdot)|_{\mathrm{LE}_p(B_R(0))}(\epsilon/3)^{-1/p})$.

Applying Proposition 7.3.2 we can find $\bar{R} > 0$ such that

$$\mathscr{L}^d\big(\Psi(t, B_R(0)) \setminus B_{\bar{R}}(0)\big) \leq \frac{\epsilon}{3C}\,,$$

where C is the constant in (a). Now, select $K_2 \subset B_{\bar{R}}(0)$ such that

- $\mathscr{L}^d\big(B_{\bar{R}}(0) \setminus K_2\big) \leq \epsilon/3C$,
- $\mathrm{Lip}(\bar{u}|_{K_2}) \leq \exp(|\bar{u}|_{\mathrm{LE}_p(B_{C(R)}(0))}(\epsilon/3C)^{-1/p})$,

where again C is as in (a). Next consider $K = K_1 \cap (\Psi(t, \cdot))^{-1}(K_2) = K_1 \cap X(t, K_2)$. Since

$$B_R(0) \setminus K \subset \big(B_R(0) \setminus K_1\big) \cup \big(B_R(0) \setminus X(t, K_2)\big)$$
$$\subset \big(B_R(0) \setminus K_1\big) \cup X\big(t, \Psi(t, B_R(0)) \setminus B_{\bar{R}}(0)\big)$$
$$\cup X(t, B_{\bar{R}}(0) \setminus K_2)\,,$$

we have

$$\mathscr{L}^d\big(B_R(0)\setminus K\big) \leq \mathscr{L}^d\big(B_R(0)\setminus K_1\big) + \mathscr{L}^d\big(X\big(t,\Psi(t,B_R(0))\setminus B_{\bar{R}}(0)\big)\big)$$
$$+ \mathscr{L}^d\big(X(t,B_{\bar{R}}(0)\setminus K_2)\big) \leq \epsilon\,.$$

Given $x,y \in K$ we have $\Psi(t,x), \Psi(t,y) \in K_2$ and hence we can estimate

$$|u(t,x) - u(t,y)| = |\bar{u}(\Psi(t,x)) - \bar{u}(\Psi(t,y))|$$

$$\leq \mathrm{Lip}(\bar{u}|_{K_2})|\Psi(t,x) - \Psi(t,y)|$$

$$\leq \mathrm{Lip}(\bar{u}|_{K_2})\mathrm{Lip}(\Psi(t,\cdot)|_{K_1})|x-y|$$

$$= |x-y|\exp\Big\{\big[(3C)^{1/p}|\bar{u}|_{\mathrm{LE}_p(B_{\bar{R}}(0))}$$

$$+3^{1/p}|\Psi(t,\cdot)|_{\mathrm{LE}_p(B_R(0))}\big]\epsilon^{-1/p}\Big\}\,.$$

Therefore $\epsilon^{1/p}\log(\mathrm{Lip}(u(t,\cdot)|_K))$ is bounded by a constant independent of ϵ and t (but which depends on R). Taking the supremum over t and ϵ, we conclude that

$$\sup_{0\leq t\leq T}|u(t,\cdot)|_{\mathrm{LE}_p(B_R(0))} \leq C(R)\,,$$

and this concludes the proof. □

7.6. An application to a conjecture on mixing flows

In [47] the author considers a problem on mixing vector fields on the two-dimensional torus $K = \mathbb{R}^2/\mathbb{Z}^2$. In this section, we are going to show that the Lipschitz estimate of Proposition 7.3.4 gives an answer to this problem, although in the L^p setting ($p > 1$) instead of the L^1 setting considered in [47].

Fix coordinates $x = (x_1,x_2) \in [0,1[\times[0,1[$ on K and consider the set

$$A = \big\{(x_1,x_2) : 0 \leq x_2 \leq 1/2\big\} \subset K\,.$$

If $b : [0,1]\times K \to \mathbb{R}^2$ is a smooth time-dependent vector field, we denote as usual by $X(t,x)$ the flow of b and by $\Phi : K \to K$ the value of the flow at time $t = 1$. We assume that the flow is nearly incompressibile, so that for some $\kappa' > 0$ we have

$$\kappa'\mathscr{L}^2(\Omega) \leq \mathscr{L}^2\big(X(t,\Omega)\big) \leq \frac{1}{\kappa'}\mathscr{L}^2(\Omega) \tag{7.24}$$

for all $\Omega \subset K$ and all $t \in [0, 1]$. For a fixed $0 < \kappa < 1/2$, we say that Φ *mixes the set A up to scale ϵ* if for every ball $B_\epsilon(x)$ we have

$$\kappa \mathcal{L}^2(B_\epsilon(x)) \leq \mathcal{L}^2\big(B_\epsilon(x) \cap \Phi(A)\big) \leq (1 - \kappa)\mathcal{L}^2(B_\epsilon(x)).$$

Then in [47] the following conjecture is proposed:

Conjecture 7.6.1 (Bressan's mixing conjecture). Under these assumptions, there exists a constant C depending only on κ and κ' such that, if Φ mixes the set A up to scale ϵ, then

$$\int_0^1 \int_K |D_x b| \, dx \, dt \geq C|\log \epsilon| \qquad \text{for every } 0 < \epsilon < 1/4.$$

In this section we show the following result:

Theorem 7.6.2. *Let $p > 1$. Under the previous assumptions, there exists a constant C depending only on κ, κ' and p such that, if Φ mixes the set A up to scale ϵ, then*

$$\int_0^1 \|D_x b\|_{L^p(K)} \, dt \geq C|\log \epsilon| \qquad \text{for every } 0 < \epsilon < 1/4.$$

Proof. We set $M = \|D_x b\|_{L^1((0,1]; L^p(K))}$ and $A' = K \setminus A$. Applying Proposition 7.3.4, and noticing that the flow is bounded since we are on the torus, for every constant $\eta > 0$ we can find a set B with $\mathcal{L}^2(B) \leq \eta$ such that

$$\text{Lip}\left(\Phi^{-1}|_{K \setminus B}\right) \leq \exp(\beta M), \qquad (7.25)$$

where the constant β depends only on κ', η and p. Since Φ mixes the set A up to scale ϵ, for every $x \in A$ we have

$$\mathcal{L}^2\big(B_\epsilon(\Phi(x)) \cap \Phi(A')\big) \geq \kappa \mathcal{L}^2\big(B_\epsilon(\Phi(x))\big). \qquad (7.26)$$

We define

$$\tilde{A} = \{x \in A \ : \ B_\epsilon(\Phi(x)) \cap [\Phi(A') \setminus B] = \emptyset\}.$$

From this definition and from (7.26) we get that for every $x \in \tilde{A}$

$$\mathcal{L}^2\big(B_\epsilon(\Phi(x)) \cap B\big) \geq \kappa \mathcal{L}^2\big(B_\epsilon(\Phi(x))\big). \qquad (7.27)$$

From (7.27) and the Besicovitch covering theorem we deduce that for an absolute constant c we have

$$\mathcal{L}^2(\Phi(\tilde{A})) \leq \frac{c}{\kappa}\mathcal{L}^2(B) \leq \frac{c\eta}{\kappa}.$$

From the compressibility condition (7.24) we deduce

$$\mathscr{L}^2(\tilde{A}) \le \frac{c\eta}{\kappa\kappa'} \, .$$

Since, using again (7.24), we know that

$$\mathscr{L}^2\big(\Phi^{-1}(B)\big) \le \frac{\mathscr{L}^2(B)}{\kappa'} \le \frac{\eta}{\kappa'} \, ,$$

we can choose $\eta > 0$, depending on κ and κ' only, in such a way that

$$\mathscr{L}^2(\tilde{A}) + \mathscr{L}^2\big(\Phi^{-1}(B)\big) \le \frac{1}{6} \, .$$

This implies the existence of a point $\bar{x} \in A \setminus \big[\tilde{A} \cup \Phi^{-1}(B)\big]$ with dist$(\bar{x}, A') \ge 1/6$. Let $\bar{y} = \Phi(\bar{x})$. Since $\bar{x} \notin \tilde{A}$, we can find a point $\bar{z} \in B_\epsilon(\bar{y}) \cap \big[\Phi(A') \setminus B\big]$. Clearly we have $|\bar{y} - \bar{z}| \le \epsilon$ and (since $\Phi^{-1}(z) \in A'$) we also have $|\bar{x} - \Phi^{-1}(\bar{z})| \ge 1/6$.

Since $\bar{y}, \bar{z} \notin B$, we can apply (7.25) to deduce

$$\frac{1}{6} \le \epsilon \mathrm{Lip}\,\big(\Phi^{-1}|_{K \setminus B}\big) \le \epsilon \exp(\beta M) \, ,$$

where now β depends only on κ, κ' and p, since η has been fixed. This implies that

$$M = \|D_x b\|_{L^1([0,1];L^p(K))} \ge \frac{1}{\beta} \log\left(\frac{1}{6\epsilon}\right) \, .$$

Hence we can find $\epsilon_0 > 0$ such that

$$M \ge \frac{1}{2\beta}|\log \epsilon| \qquad \text{for every } 0 < \epsilon < \epsilon_0.$$

We are now going to show the thesis for every $0 < \epsilon < 1/4$. Indeed, suppose that the thesis is false. Then, we could find a sequence $\{b_h\}$ of vector fields and a sequence $\{\epsilon_h\}$ with $\epsilon_0 < \epsilon_h < 1/4$ in such a way that

$$\|D_x b_h\|_{L^1([0,1];L^p(K))} \le \frac{1}{h}|\log \epsilon_h|$$

and the corresponding map Φ_h mixes the set A up to scale ϵ_h. This implies that

$$\|D_x b_h\|_{L^1([0,1];L^p(K))} \le \frac{1}{h}|\log \epsilon_h| \le \frac{1}{h}|\log \epsilon_0| \longrightarrow 0 \qquad \text{as } h \to \infty.$$

Up to an extraction of a subsequence, we can assume that $\epsilon_h \to \bar{\epsilon}$ and that $\Phi_h \to \Phi$ strongly in $L^1(K)$. For this, we apply the compactness result in Theorem 7.3.6, noticing that (7.24) gives a uniform control on the compressibility constants of the flows and that we do not need any assumption on the growth of the vector fields, since we are on the torus and then the flow is automatically uniformly bounded. Now, notice that the mixing property is stable with respect to strong convergence: this means that Φ has to mix up to scale $\bar{\epsilon} \leq 1/4$. But since $\|D_x b_h\|_{L^1([0,1];L^p(K))} \to 0$, we deduce that Φ is indeed a translation on K, hence it cannot mix the set A up to a scale which is smaller than $1/4$. From this contradiction we get the thesis. $\square$

Remark 7.6.3. We notice that the constant $1/4$ in Theorem 7.6.2 depends on the shape of the set A: this bound comes from the fact that a translation does not mix up to a scale $\epsilon < 1/4$. Our proof can be easily extended to the case of a measurable set A with any shape, giving a different upper bound for the values of ϵ such that the result is true.

Appendix A
Background material and technical results

In this appendix we collect some results on measure theory, Lipschitz functions, BV functions (for which good general references are [17, 18, 91, 92, 96]) and on the theory of maximal functions (for which we refer to [120]).

A.1. Measure theory

Let X be a separable metric space. We denote by $\mathscr{B}(X)$ the family of the Borel subsets of X, by $\mathcal{P}(X)$ the family of the Borel probability measures on X, by $\mathcal{M}(X)$ the family of the locally finite Borel measures on X and by $\mathcal{M}_+(X)$ the subset of $\mathcal{M}(X)$ consisting of all nonnegative locally finite Borel measures on X. In a similar fashion we can consider vector-valued or matrix-valued measures on X. The *total variation* of a measure $\mu \in \mathcal{M}(X)$ is denoted by $|\mu|$. In the case when $|\mu|(X) < +\infty$ we say that μ has finite mass and we set $\|\mu\|_{\mathcal{M}(X)} = |\mu|(X)$. The *support* of a measure $\mu \in \mathcal{M}(X)$ is the closed set defined by

$$\operatorname{spt}\mu = \{x \in X \ : \ |\mu|(U) > 0 \text{ for each open neighborhood } U \text{ of } x \} \ .$$

A measure $\mu \in \mathcal{M}(X)$ is *concentrated* on a Borel set $E \subset X$ if it satisfies $|\mu|(X \setminus E) = 0$.

If $\mu \in \mathcal{M}(X)$ and $\nu \in \mathcal{M}_+(X)$ we say that μ is *absolutely continuous* with respect to ν (and we write $\mu \ll \nu$) if $|\mu|(E) = 0$ for every Borel set $E \subset X$ such that $\nu(E) = 0$. We say that two measures μ and $\nu \in \mathcal{M}(X)$ are *mutually singular* (and we write $\mu \perp \nu$) if they are concentrated on disjoint Borel sets. If $\mu \in \mathcal{M}(X)$ and $E \subset X$ is a Borel set, the *restriction* of μ to E is the measure $\mu \llcorner E \in \mathcal{M}(X)$ defined *via* the formula $(\mu \llcorner E)(A) = \mu(A \cap E)$ for every Borel set $A \subset X$. If $f : X \to Y$ is a Borel map between two separable metric spaces X and Y and $\mu \in \mathcal{M}(X)$ we denote by $f_\# \mu \in \mathcal{M}(Y)$ the *push-forward* of the measure μ, defined by

$$(f_\# \mu)(E) = \mu(f^{-1}(E)) \qquad \text{for every Borel set } E \subset Y. \qquad (A.1)$$

We denote by $\mathscr{L}^d$ the Lebesgue measure on $\mathbb{R}^d$ and with $\mathscr{H}^k$ the k-dimensional Hausdorff measure on $\mathbb{R}^d$. For any subset A of $\mathbb{R}^d$ we define the characteristic function $\mathbf{1}_A$ as

$$\mathbf{1}_A(x) = \begin{cases} 1 & \text{if } x \in A \\ 0 & \text{if } x \notin A. \end{cases}$$

A family $\mathscr{F} \subset \mathcal{M}_+(X)$ is *bounded* if there exists a constant C with $\|\mu\|_{\mathcal{M}(X)} \leq C$ for each $\mu \in \mathscr{F}$. We say that a bounded sequence $\{\mu_h\} \subset \mathcal{M}_+(X)$ is *narrowly convergent* to $\mu \in \mathcal{M}_+(X)$ as $h \to \infty$ if

$$\lim_{h \to \infty} \int_X f(x)\, d\mu_h(x) = \int_X f(x)\, d\mu(x)$$

for every $f \in C_b(X)$, the space of continuous and bounded real functions defined on X. We say that a bounded family $\mathscr{F} \subset \mathcal{M}_+(X)$ is *tight* if for every $\epsilon > 0$ there exists a compact set $K_\epsilon \subset X$ such that $\mu(X \setminus K_\epsilon) \leq \epsilon$ for every $\mu \in \mathscr{F}$. The following theorem characterizes the relatively compact subsets with respect to the narrow topology.

Theorem A.1.1 (Prokhorov). *Assume that X is a complete metric space. Then a bounded family $\mathscr{F} \subset \mathcal{M}_+(X)$ is relatively compact with respect to the narrow convergence if and only if it is tight.*

Moreover, a necessary and sufficient condition for tightness is the existence of a coercive functional $\Psi : X \to [0, +\infty]$ such that $\int_X \Psi\, d\mu \leq 1$ for every $\mu \in \mathscr{F}$. It is also simple to check that a bounded family $\mathscr{F} \subset \mathcal{M}_+(X \times Y)$ is tight if and only if the families of the marginals $(\pi_X)_{\#}\mathscr{F} \subset \mathcal{M}_+(X)$ and $(\pi_Y)_{\#}\mathscr{F} \subset \mathcal{M}_+(Y)$ are tight.

We also recall a particular case of the *disintegration theorem*. Let $\mu \in \mathcal{M}_+(X \times Y)$, define $\nu = (\pi_X)_{\#}\mu$ and assume that $\nu \in \mathcal{M}_+(X)$. Then there exists a Borel family $\{\mu_x\}_{x \in X} \subset \mathcal{M}_+(Y)$, which is uniquely determined ν-a.e., such that

$$\mu = \int_X \mu_x\, d\nu(x).$$

Fix a measure $\mu \in \mathcal{M}_+(X)$. We say that a sequence of functions $\{f_h\}$ defined on X and with values in $\mathbb{R}^k$ *converges in μ-measure* to a function f if

$$\lim_{h \to \infty} \mu(\{x \in X \ : \ |f_h(x) - f(x)| > \delta\}) = 0 \qquad \text{for every } \delta > 0.$$

We say that a Borel set $\Sigma \subset \mathbb{R}^d$ is $\mathscr{H}^k$-*rectifiable* if there exist countably many Lipschitz functions $f_i : \mathbb{R}^k \to \mathbb{R}^d$ such that

$$\mathscr{H}^k\left(\Sigma \setminus \bigcup_i f_i(\mathbb{R}^k)\right) = 0.$$

We finally recall the *coarea formula* for Lipschitz functions (see for instance [17, Section 2.12]): for every Lipschitz map $f : \mathbb{R}^d \to \mathbb{R}^{d-k}$ and every positive Borel function $\varphi : \mathbb{R}^d \to [0, +\infty]$ there holds

$$\int_{\mathbb{R}^d} \varphi J \, d\mathscr{L}^d = \int_{\mathbb{R}^{d-k}} \left[\int_{E_h} \varphi \, d\mathscr{H}^k \right] d\mathscr{L}^{d-k}(h), \qquad (A.2)$$

where $J = [\det(\nabla f \cdot {}^t \nabla f)]^{1/2}$ is the Jacobian of f and $E_h = \{x \in \mathbb{R}^d : f(x) = h\}$, for $h \in \mathbb{R}^{d-k}$, are the level sets of the function f. If $\phi \in L^1(\mathbb{R}^d)$ from (A.2) we also deduce

$$\int_{\mathbb{R}^d} \phi \, d\mathscr{L}^d = \int_{\mathbb{R}^{d-k}} \left[\int_{E_h} \frac{\phi}{J} \, d\mathscr{H}^k \right] d\mathscr{L}^{d-k}(h). \qquad (A.3)$$

A.2. Lipschitz functions, extension theorems and approximate differentiability

A map $f : \Omega \subset \mathbb{R}^d \to \mathbb{R}^k$ is *Lipschitz* if there exists a constant L such that

$$|f(x) - f(y)| \leq L|x - y| \qquad \text{for every } x, y \in \Omega. \qquad (A.4)$$

The minimal constant L such that (A.4) holds is called the *Lipschitz constant* of the function f and is denoted by $\mathrm{Lip}(f)$.

Theorem A.2.1 (Rademacher). *A Lipschitz function* $f : \mathbb{R}^d \to \mathbb{R}^k$ *is differentiable at* $\mathscr{L}^d$*-a.e.* $x \in \mathbb{R}^d$.

It is simple to check that every real-valued Lipschitz function $f : \Omega \subset \mathbb{R}^d \to \mathbb{R}$ can be extended to a function $\tilde{f} : \mathbb{R}^d \to \mathbb{R}$ with $\mathrm{Lip}(\tilde{f}) = \mathrm{Lip}(f)$ (by extension we mean that $\tilde{f}|_\Omega = f$). Indeed, it is sufficient to set $\tilde{f} = f^+$ or $\tilde{f} = f^-$, where

$$f^{\pm}(x) = \inf \left\{ f(y) \pm L|x - y| \; : \; y \in \Omega \right\}$$

and $L = \mathrm{Lip}(f)$. Let us notice that f^+ and f^- are respectively the biggest and the smallest extensions. The same result holds also if Ω is a metric space. Moreover, arguing componentwise, we deduce that every Lipschitz function $f : \Omega \subset \mathbb{R}^d \to \mathbb{R}^k$ can be extended to a function $\tilde{f} : \mathbb{R}^d \to \mathbb{R}^k$ with $\mathrm{Lip}(\tilde{f}) = \sqrt{k} \, \mathrm{Lip}(f)$. In fact, for subsets of $\mathbb{R}^d$, also the following stronger result holds:

Theorem A.2.2 (Kirszbraun). *Let* $f : \Omega \subset \mathbb{R}^d \to \mathbb{R}^k$ *be a Lipschitz function. Then there exists an extension* $\tilde{f} : \mathbb{R}^d \to \mathbb{R}^k$ *with* $\mathrm{Lip}(\tilde{f}) = \mathrm{Lip}(f)$.

Its proof is considerably more difficult and we mention it for completeness, although it is not strictly necessary in our presentation.

We say that a Borel map $f : \mathbb{R}^d \to \mathbb{R}^k$ is *approximately differentiable* at $x \in \mathbb{R}^d$ if there exists a linear map $L : \mathbb{R}^d \to \mathbb{R}^k$ such that the difference quotients

$$y \mapsto \frac{f(x + \epsilon y) - f(x)}{\epsilon}$$

locally converge in measure as $\epsilon \downarrow 0$ to Ly. This is clearly a local property. It is possible to show that, if $f|_K$ is a Lipschitz map for some set $K \subset \mathbb{R}^d$, then f is approximately differentiable at almost every point of K. In the following theorem we show a kind of converse of this statement: an approximately differentiable map can be approximated, in the Lusin sense, with Lipschitz maps.

Theorem A.2.3. *Let $f : \Omega \to \mathbb{R}^k$. Assume that there exists a sequence of Borel sets $A_h \subset \Omega$ such that $\mathscr{L}^d\left(\Omega \setminus \bigcup_h A_h\right) = 0$ and $f|_{A_h}$ is Lipschitz for any h. Then f is approximately differentiable at $\mathscr{L}^d$-a.e. $x \in \Omega$. Conversely, if f is approximately differentiable at all points of $\Omega' \subset \Omega$, we can write Ω' as a countable union of sets A_h such that $f|_{A_h}$ is Lipschitz for any h (up to a redefinition of f in an $\mathscr{L}^d$-negligible set).*

A.3. Functions with bounded variation

We say that a function $f : \Omega \subset \mathbb{R}^d \to \mathbb{R}$ has *bounded variation*, and we write $f \in BV(\Omega)$, if $f \in L^1(\Omega)$ and the distributional derivative Df is a vector-valued finite measure in Ω. We also introduce the space $BV_{\text{loc}}(\Omega)$ of functions with *locally bounded variation* as the class of those $f : \Omega \subset \mathbb{R}^d \to \mathbb{R}$ such that $f \in BV(\tilde{\Omega})$ for every $\tilde{\Omega} \Subset \Omega$. The spaces $BV(\Omega; \mathbb{R}^k)$ and $BV_{\text{loc}}(\Omega; \mathbb{R}^k)$ are defined by requiring BV or BV_{loc} regularity on each component. The space $BV(\Omega)$ is a Banach space with the norm

$$\|f\|_{BV(\Omega)} = \|f\|_{L^1(\Omega)} + \|Df\|_{\mathcal{M}(\Omega)} .$$

When $f \in L^1(\Omega; \mathbb{R}^k)$ we can consider the set of points of *approximate discontinuity* S_f defined by

$$\Omega \setminus S_f = \left\{ x \in \Omega \ : \ \exists z \in \mathbb{R}^k \text{ s.t. } \fint_{B_r(x)} |f(y) - z| \, dy \to 0 \text{ as } r \to 0 \right\} .$$

Every $x \in \Omega \setminus S_f$ is called a point of *approximate continuity*. The value $z \in \mathbb{R}^k$ appearing in the definition, when it exists, is unique and is denoted by $\tilde{f}(x)$. It is simple to check that S_f is a Borel set and that the equality

$f = \tilde{f}$ holds $\mathcal{L}^d$-a.e. in $\Omega \setminus S_f$. We also introduce the set of points of *approximate jump* $J_f \subset S_f$:

$$J_f = \left\{ x \in S_f : \exists v \in \mathbb{S}^{d-1}, \exists z^{\pm} \in \mathbb{R}^k \text{ s.t.} \fint_{B_{v,r}^{\pm}(x)} |f(y) - z^{\pm}| dy \to 0 \text{ as } r \to 0 \right\},$$

where we have defined the half-balls

$$B_{v,r}^{\pm}(x) = \{ y \in B_r(x) : (y - x) \cdot v \gtrless 0 \} .$$

The triple (v, z^-, z^+), when it exists, is unique up to a permutation of z^- and z^+ and a change of sign of v, and is denoted by $(v(x), f^-(x), f^+(x))$. Notice that $f^-(x) \neq f^+(x)$ when they exist, since $x \in S_f$. It is easy to check that J_f is a Borel set and that $f^{\pm}$ and v can be chosen to be Borel functions in their domain of definition.

For $f \in BV(\Omega; \mathbb{R}^k)$, with $\Omega \subset \mathbb{R}^d$, we can use the Lebesgue decomposition theorem to obtain

$$Df = D^a f + D^c f ,$$

with $D^a f \ll \mathcal{L}^d$ and $D^s \perp \mathcal{L}^d$. The measure $D^a f$ is called the *absolutely continuous* part of the derivative and the measure $D^s f$ is called the *singular part* of the derivative. The singular part can be decomposed as

$$D^s f = D^j f + D^c f = D^s f \llcorner J_f + D^s f \llcorner (\Omega \setminus J_f) ,$$

where $D^j f$ is the *jump* part of the derivative and $D^c f$ is the *Cantor* part of the derivative. The subspace $SBV(\Omega; \mathbb{R}^k) \subset BV(\Omega; \mathbb{R}^k)$ of *special function with bounded variation* consists of those $f \in BV(\Omega; \mathbb{R}^k)$ with $D^c f = 0$.

The following important result relative to the structure of BV functions holds. For the notion of rectifiable set we refer to Appendix A.1.

Theorem A.3.1 (Structure of BV functions). *Let $f \in BV(\Omega; \mathbb{R}^k)$. Then the approximate jump set is countably $\mathcal{H}^{d-1}$-rectifiable, $\mathcal{H}^{d-1}(S_f \setminus J_f) = 0$ and*

$$D^j f = (f^+ - f^-) \otimes v \, \mathcal{H}^{d-1} \llcorner J_f . \tag{A.5}$$

The following deep result by Alberti (see [3] and the recent simplified presentation of [78]) says that the rank-one structure of the jump part of the derivative expressed by (A.5) is also shared by the Cantor part.

Theorem A.3.2 (Alberti's Rank-one theorem). *Let $f \in BV(\Omega; \mathbb{R}^k)$. Then there exist two Borel functions $\xi : \Omega \to \mathbb{S}^{d-1}$ and $\eta : \Omega \to \mathbb{S}^{k-1}$ such that*

$$D^s f = \xi \otimes \eta |D^s f| .$$

In a similar fashion we can define the space of vector fields with *bounded deformation* in an open set $\Omega \subset \mathbb{R}^d$ by requiring that the simmetric part of the distributional derivative $Ef = \frac{1}{2}(Df + {}^t Df)$ is a measure:

$$BD(\Omega; \mathbb{R}^d) = \{ f \in L^1(\Omega; \mathbb{R}^d) : Ef \text{ is a matrix valued finite measure in } \Omega \}.$$

We refer to [123, 12] for a presentation of the main properties of this class of vector fields. As in the BV case we can decompose $Ef = E^a f + E^j f + E^c f$ and define the space of *special vector fields with bounded deformation* as

$$SBD(\Omega; \mathbb{R}^d) = \{ f \in BD(\Omega; \mathbb{R}^d) \ : \ E^c f = 0 \} \ .$$

A.4. Maximal functions

We recall here the definition of the *local maximal function* of a locally finite measure and of a locally summable function and we state some well-known properties.

Definition A.4.1 (Local maximal function). Let μ be a (vector-valued) locally finite measure. For every $\lambda > 0$, we define the *local maximal function* of μ as

$$M_\lambda \mu(x) \ = \ \sup_{0 < r < \lambda} \frac{|\mu|(B_r(x))}{\mathscr{L}^d(B_r(x))} = \sup_{0 < r < \lambda} \fint_{B_r(x)} d|\mu|(y) \qquad x \in \mathbb{R}^d \ .$$

When $\mu = f\mathscr{L}^d$, where f is a function in $L^1_{\text{loc}}(\mathbb{R}^d; \mathbb{R}^k)$, we will often use the notation $M_\lambda f$ for $M_\lambda \mu$.

In Chapter 7 we made an extensive use of the following two lemmas.

Lemma A.4.2. *Let $\lambda > 0$. The local maximal function of μ is finite for $\mathscr{L}^d$-a.e. $x \in \mathbb{R}^d$ and we have*

$$\int_{B_\rho(0)} M_\lambda f(y) \, dy \leq c_{d,\rho} + c_d \int_{B_{\rho+\lambda}(0)} |f(y)| \log(2 + |f(y)|) \, dy \ .$$

For $p > 1$ and $\rho > 0$ we have the strong *estimate*

$$\int_{B_\rho(0)} (M_\lambda f(y))^p \, dy \leq c_{d,p} \int_{B_{\rho+\lambda}(0)} |f(y)|^p \, dy \ ,$$

which is however false for $p = 1$. For $p = 1$ we only have the weak *estimate*

$$\mathscr{L}^d \left(\{ y \in B_\rho(0) \ : \ M_\lambda f(y) > t \} \right) \leq \frac{c_d}{t} \int_{B_{\rho+\lambda}(0)} |f(y)| \, dy \ , \qquad \text{(A.6)}$$

for every $t > 0$.

Lemma A.4.3. *If $u \in BV(\mathbb{R}^d)$ then there exists an $\mathscr{L}^d$-negligible set $N \subset \mathbb{R}^d$ such that*

$$|u(x) - u(y)| \leq c_d |x - y| \left(M_\lambda Du(x) + M_\lambda Du(y) \right)$$

for $x, y \in \mathbb{R}^d \setminus N$ with $|x - y| \leq \lambda$.

We also recall the Chebyshev inequality: for every $t > 0$ we have

$$\mathscr{L}^d(\{|f| > t\}) \leq \frac{1}{t} \int_{\{|f|>t\}} |f(x)| \, dx \leq \frac{\mathscr{L}^d(\{|f| > t\})^{1/q}}{t} \|f\|_{L^p(\Omega)} \, ,$$

where $\frac{1}{p} + \frac{1}{q} = 1$, and this implies

$$\mathscr{L}^d(\{|f| > t\})^{1/p} \leq \frac{\|f\|_{L^p(\Omega)}}{t} \, . \tag{A.7}$$

References

[1] R. A. ADAMS, *Sobolev spaces*, Pure and Applied Mathematics **65** (1975), Academic Press, New York–London.

[2] M. AIZENMAN, *On vector fields as generators of flows: a counterexample to Nelson's conjecture*, Ann. Math. **107** (1978), 287–296.

[3] G. ALBERTI, *Rank-one properties for derivatives of functions with bounded variation*, Proc. Roy. Soc. Edinburgh Sect. A **123** (1993), 239–274.

[4] G. ALBERTI and L. AMBROSIO, *A geometric approach to monotone functions in* $\mathbb{R}^n$, Math. Z. **230** (1999), 259–316.

[5] G. ALBERTI, S. BIANCHINI and G. CRIPPA, Work in preparation.

[6] G. ALBERTI and S. MÜLLER, *A new approach to variational problems with multiple scales*, Comm. Pure Appl. Math. **54** (2001), 761–825.

[7] F. J. ALMGREN, *The theory of varifolds – A variational calculus in the large*, Princeton University Press, 1972.

[8] L. AMBROSIO, *Transport equation and Cauchy problem for BV vector fields*, Invent. Math. **158** (2004), 227–260.

[9] L. AMBROSIO, *Lecture notes on transport equation and Cauchy problem for BV vector fields and applications*, Preprint, 2004 (available at http://cvgmt.sns.it).

[10] L. AMBROSIO, *Transport equation and Cauchy problem for non-smooth vector fields and applications*, Lecture Notes in Mathematics "Calculus of Variations and Non-Linear Partial Differential Equation" (CIME Series, Cetraro, 2005) **1927**, B. Dacorogna and P. Marcellini eds., 2–41, 2008.

[11] L. AMBROSIO, F. BOUCHUT and C. DE LELLIS, *Well-posedness for a class of hyperbolic systems of conservation laws in several space dimensions*, Comm. PDE **29** (2004), 1635–1651.

[12] L. AMBROSIO, A. COSCIA and G. DAL MASO, *Fine properties of functions with bounded deformation*, Arch. Rational Mech. Anal. **139** (1997), 201–238.

[13] L. AMBROSIO and G. CRIPPA, *Existence, uniqueness, stability and differentiability properties of the flow associated to weakly differentiable vector fields*, In: Transport Equations and Multi-D Hyperbolic Conservation Laws, Lecture Notes of the Unione Matematica Italiana, Vol. 5, Springer (2008).

[14] L. AMBROSIO, G. CRIPPA and S. MANIGLIA, *Traces and fine properties of a BD class of vector fields and applications*, Ann. Sci. Toulouse **XIV** (4) (2005), 527–561.

[15] L. AMBROSIO and C. DE LELLIS, *Existence of solutions for a class of hyperbolic systems of conservation laws in several space dimensions*, International Mathematical Research Notices **41** (2003), 2205–2220.

[16] L. AMBROSIO, C. DE LELLIS and J. MALÝ: *On the chain-rule for the divergence of BV like vector fields: applications, partial results, open problems*, Perspectives in nonlinear partial differential equations, 31–67, Contemp. Math. **446**, Amer. Math. Soc., Providence, RI, (2007).

[17] L. AMBROSIO, N. FUSCO and D. PALLARA, *Functions of bounded variation and free discontinuity problems*, Oxford Mathematical Monographs, 2000.

[18] L. AMBROSIO, N. GIGLI and G. SAVARÉ, *Gradient flows in metric spaces and in the Wasserstein space of probability measures*, Lectures in Mathematics, ETH Zurich, Birkhäuser, 2005.

[19] L. AMBROSIO, M. LECUMBERRY and S. MANIGLIA: *Lipschitz regularity and approximate differentiability of the DiPerna–Lions flow*, Rend. Sem. Mat. Univ. Padova **114** (2005), 29–50.

[20] L. AMBROSIO, S. LISINI and G. SAVARÉ, *Stability of flows associated to gradient vector fields and convergence of iterated transport maps*, Manuscripta Math. **121** (2006), 1–50.

[21] L. AMBROSIO and J. MALÝ, *Very weak notions of differentiability*, Proceedings of the Royal Society of Edinburgh **137A** (2007), 447–455.

[22] L. AMBROSIO, P. TILLI and L. ZAMBOTTI, *Introduzione alla Teoria della Misura ed alla Probabilità*, Lecture notes of a course given at the Scuola Normale Superiore, unpublished.

[23] G. ANZELLOTTI, *Pairings between measures and bounded functions and compensated compactness*, Ann. Mat. Pura App. **135** (1983), 293–318.

[24] G. ANZELLOTTI, *The Euler equation for functionals with linear growth*, Trans. Amer. Mat. Soc. **290** (1985), 483–501.

[25] G. ANZELLOTTI, *Traces of bounded vectorfields and the divergence theorem*, Unpublished preprint, 1983.

[26] H. BAHOURI and J.-Y. CHEMIN, *Equations de transport relatives à des champs de vecteurs non-lipschitziens et mécanique des fluides*, Arch. Rational Mech. Anal. **127** (1994), 159–181.

[27] E. J. BALDER, *New fundamentals of Young measure convergence*, CRC Res. Notes in Math. **411**, 2001.

[28] J. BALL and R. JAMES, *Fine phase mixtures as minimizers of energy*, Arch. Rational Mech. Anal. **100** (1987), 13–52.

[29] V. BANGERT, *Minimal measures and minimizing closed normal one-currents*, Geom. funct. anal. **9** (1999), 413–427.

[30] A. BECK, *Uniqueness of flow solutions of differential equations*, Recent advances in topological dynamics. Lecture Notes in Math., Vol. 318, Springer, Berlin, 1973.

[31] J.-D. BENAMOU and Y. BRENIER, *Weak solutions for the semi-geostrophic equation formulated as a couples Monge-Ampere transport problem*, SIAM J. Appl. Math. **58** (1998), 1450–1461.

[32] P. BERNARD and B. BUFFONI, *Optimal mass transportation and Mather theory*, J. Eur. Math. Soc. (JEMS) **9** (2007), 85–121.

[33] O. BESSON and J. POUSIN, *Solutions for linear conservation laws with velocity fields in L^∞*, Arch. Rational Mech. Anal. **186** (2007), 159–175.

[34] S. BIANCHINI and A. BRESSAN, *Vanishing viscosity solutions of nonlinear hyperbolic systems*, Ann. Math. **161** (2005), 223–342.

[35] V. BOGACHEV and E. M. WOLF, *Absolutely continuous flows generated by Sobolev class vector fields in finite and infinite dimensions*, J. Funct. Anal. **167** (1999), 1–68.

[36] F. BOUCHUT, *Renormalized solutions to the Vlasov equation with coefficients of bounded variation*, Arch. Rational Mech. Anal. **157** (2001), 75–90.

[37] F. BOUCHUT and G. CRIPPA, *Uniqueness, Renormalization, and Smooth Approximations for Linear Transport Equations*, SIAM J. Math. Anal. **38** (2006), 1316–1328.

[38] F. BOUCHUT and L. DESVILLETTES, *On two-dimensional Hamiltonian transport equations with continuous coefficients*, Differential Integral Equations **14** (2001), 1015–1024.

[39] F. BOUCHUT, F. GOLSE and M. PULVIRENTI, *Kinetic equations and asymptotic theory*, Series in Appl. Math., Gauthiers-Villars, 2000.

[40] F. BOUCHUT and F. JAMES, *One-dimensional transport equation with discontinuous coefficients*, Nonlinear Analysis **32** (1998), 891–933.

[41] F. BOUCHUT, F. JAMES and S. MANCINI, *Uniqueness and weak stability for multi-dimensional transport equations with one-sided Lipschitz coefficients*, Annali Scuola Normale Superiore, Ser. 5 **4** (2005), 1–25.

[42] Y. BRENIER, *The least action principle and the related concept of generalized flows for incompressible perfect fluids*, J. Amer. Mat. Soc. **2** (1989), 225–255.

[43] Y. BRENIER, *The dual least action problem for an ideal, incompressible fluid*, Arch. Rational Mech. Anal. **122** (1993), 323–351.

[44] Y. BRENIER, *A homogenized model for vortex sheets*, Arch. Rational Mech. Anal. **138** (1997), 319–353.

[45] Y. BRENIER, *Minimal geodesics on groups of volume-preserving maps and generalized solutions of the Euler equations*, Comm. Pure Appl. Math. **52** (1999), 411–452.

[46] A. BRESSAN, *Hyperbolic systems of conservation laws. The one-dimensional Cauchy problem*, Oxford Lecture Series in Mathematics and its Applications **20**. Oxford University Press, Oxford, 2000.

[47] A. BRESSAN, *A lemma and a conjecture on the cost of rearrangements*, Rend. Sem. Mat. Univ. Padova **110** (2003), 97–102.

[48] A. BRESSAN, *An ill posed Cauchy problem for a hyperbolic system in two space dimensions*, Rend. Sem. Mat. Univ. Padova **110** (2003), 103–117.

[49] H. BREZIS, *Analyse fonctionnelle. Théorie et applications*, Masson, Paris, 1983.

[50] I. CAPUZZO DOLCETTA and B. PERTHAME, *On some analogy between different approaches to first order PDE's with nonsmooth coefficients*, Adv. Math. Sci Appl. **6** (1996), 689–703.

[51] A. CELLINA, *On uniqueness almost everywhere for monotonic differential inclusions*, Nonlinear Analysis, TMA **25** (1995), 899–903.

[52] A. CELLINA and M. VORNICESCU, *On gradient flows*, Journal of Differential Equations **145** (1998), 489–501.

[53] J.-Y. CHEMIN, *Perfect incompressible fluids*, Oxford University Press, 1998.

[54] J.-Y. CHEMIN and N. LERNER, *Flot de champs de vecteurs non lipschitziens et équations de Navier–Stokes*, Journal of Differential Equations **121** (1995), 314–328.

[55] G.-Q. CHEN and H. FRID, *Divergence-measure fields and conservation laws*, Arch. Rational Mech. Anal. **147** (1999), 89–118.

[56] G.-Q. CHEN and H. FRID, *Extended divergence-measure fields and the Euler equation of gas dynamics*, Comm. Math. Phys. **236** (2003), 251–280.

[57] F. COLOMBINI, G. CRIPPA and J. RAUCH, *A note on two-dimensional transport with bounded divergence*, Comm. PDE **31** (2006), 1109–1115.

[58] F. COLOMBINI and N. LERNER, *Sur les champs de vecteurs peu réguliers*, Séminaire: Équations aux Dérivées Partielles, Exp. No. XIV, École Polytech., Palaiseau, 2001.

[59] F. COLOMBINI and N. LERNER, *Uniqueness of continuous solutions for BV vector fields*, Duke Math. J. **111** (2002), 357–384.

[60] F. COLOMBINI and N. LERNER, *Uniqueness of L^∞ solutions for a class of conormal BV vector fields*, Contemp. Math. **368** (2005), 133–156.

[61] F. COLOMBINI, T. LUO and J. RAUCH, *Uniqueness and non-uniqueness for nonsmooth divergence free transport*, Séminaire: Équations aux Dérivées Partielles, Exp. No. XXII, École Polytech., Palaiseau, 2003.

[62] F. COLOMBINI, T. LUO and J. RAUCH, *Nearly Lipschitzean divergence-free transport propagates neither continuity nor BV regularity*, Commun. Math. Sci. **2** (2004), 207–212.

[63] F. COLOMBINI and J. RAUCH, *Uniqueness in the Cauchy Problem for Transport in $\mathbb{R}^2$ and $\mathbb{R}^{1+2}$*, J. Differential Equations **211** (2005), 162–167.

[64] G. CRIPPA, *Equazione del trasporto e problema di Cauchy per campi vettoriali debolmente differenziabili*, Tesi di Laurea, Università di Pisa, 2006 (available at http://cvgmt.sns.it).

[65] G. CRIPPA and C. DE LELLIS, *Oscillatory solutions to transport equations*, Indiana Univ. Math. J. **55** (2006), 1–13.

[66] G. CRIPPA and C. DE LELLIS, *Estimates and regularity results for the DiPerna–Lions flow*, J. Reine Angew. Math. **616** (2008), 15–46.

[67] G. CRIPPA and C. DE LELLIS, *Regularity and compactness for the DiPerna–Lions flow*, In: Hyperbolic Problems: Theory, Numerics, Applications, Proceedings of the conference HYP2006 (Lyon, July 17-21, 2006).

[68] A. B. CRUZEIRO, *Équations différentielles ordinaires: non explosion et mesures quasi-invariantes*, J. Funct. Anal. **54** (1983), 193–205.

[69] A. B. CRUZEIRO, *Équations différentielles sur l'espace de Wiener et formules de Cameron-Martin non linéaires*, J. Funct. Anal. **54** (1983), 206–227.

[70] A. B. CRUZEIRO, *Unicité de solutions d'équations différentielles sur l'espace de Wiener*, J. Funct. Anal. **58** (1984), 335–347.

[71] M. CULLEN, *On the accuracy of the semi-geostrophic approximation*, Quart. J. Roy. Metereol. Soc. **126** (2000), 1099–1115.

[72] M. CULLEN and M. FELDMAN, *Lagrangian solutions of semi-geostrophic equations in physical space*, SIAM J. Math. Anal. **37** (2006), 1371–1395.

[73] M. CULLEN and W. GANGBO, *A variational approach for the two-dimensional semi-geostrophic shallow water equations*, Arch. Rational Mech. Anal. **156** (2001), 241–273.

[74] C. DAFERMOS, *Polygonal approximations of solutions of the initial value problem for a conservation law*, J. Math. Anal. Appl. **38** (1972), 33–41.

[75] C. DAFERMOS, *Hyperbolic conservation laws in continuum physics*, Springer Verlag, 2000.

[76] C. DE LELLIS, *Blow-up of the BV norm in the multi-dimensional Keyfitz and Kranzer system*, Duke Math. J. **127** (2004), 313–339.

[77] C. DE LELLIS, *Notes on hyperbolic systems of conservation laws and transport equations*, Handbook of Differential Equations: Evolutionary Equations, vol. III. Edited by C. M. Dafermos and E. Feireisl. Elsevier/North-Holland, Amsterdam, 2006.

[78] C. DE LELLIS, *A note on Alberti's Rank-one Theorem*, In: Transport Equations and Multi-D Hyperbolic Conservation Laws, Lecture Notes of the Unione Matematica Italiana, Vol. 5, Springer (2008).

[79] C. DE LELLIS, *Ordinary differential equations with rough coefficients and the renormalization theorem of Ambrosio (d'après Ambrosio, DiPerna, Lions)*, Séminaire Bourbaki, vol. 2006/2007, n. 972.

[80] C. DE LELLIS and M. WESTDICKENBERG, *On the optimality of velocity averaging lemmas*, Ann. Inst. H. Poincaré Anal. Non Linéaire **20** (2003), 1075–1085.

[81] L. DE PASCALE, M. S. GELLI and L. GRANIERI, *Minimal measures, one-dimensional currents and the Monge-Kantorovich problem*, Calc. Var. Partial Differential Equations **27** (2006), 1–23.

[82] N. DEPAUW, *Non unicité des solutions bornées pour un champ de vecteurs BV en dehors d'un hyperplan*, C.R. Math. Sci. Acad. Paris **337** (2003), 249–252.

[83] R. J. DIPERNA, *Measure-valued solutions to conservation laws*, Arch. Rational Mech. Anal. **88** (1985), 223–270.

[84] R. J. DiPERNA and P.-L. LIONS, *Ordinary differential equations, transport theory and Sobolev spaces*, Invent. Math. **98** (1989), 511–547.

[85] R. J. DiPERNA and P.-L. LIONS, *On the Cauchy problem for the Boltzmann equation: global existence and weak stability*, Ann. of Math. **130** (1989), 312–366.

[86] R. ENGELKING, *General Topology. Revised and completed edition*, Sigma series in pure mathematics **6**. Heldermann Verlag, Berlin (1989).

[87] L. C. EVANS, *Partial Differential Equations and Monge–Kantorovich Mass Transfer*, Current Developments in Mathematics (1997), 65–126.

[88] L. C. EVANS, *Partial Differential Equations*, Graduate studies in Mathematics **19** (1998), American Mathematical Society.

[89] L. C. EVANS and W. GANGBO, *Differential equations methods for the Monge-Kantorovich mass transfer problem*, Mem. Amer. Math. Soc. **137** (1999), no. 653.

[90] L. C. EVANS, W. GANGBO and O. SAVIN, *Diffeomorphisms and nonlinear heat flows*, SIAM J. Math. Anal. **37** (2005), 737–751.

[91] L. C. EVANS and R. F. GARIEPY, *Lecture notes on measure theory and fine properties of functions*, CRC Press, 1992.

[92] H. FEDERER, *Geometric measure theory*, Springer, 1969.

[93] A. FIGALLI, *Existence and uniqueness for SDEs with rough or degenerate coefficients*, J. Funct. Anal. **254** (2008), no. 1, 109–153.

[94] M. GIAQUINTA and G. MODICA, *Analisi matematica III: Strutture lineari e metriche, continuità*, Pitagora Editrice, Bologna. 2000.

[95] D. GILBARG and N. S. TRUDINGER, *Elliptic partial differential equations of second order*, Classics in Mathematics. Springer-Verlag, Berlin, 2001.

[96] E. GIUSTI, *Minimal surfaces and functions of bounded variation*, Monographs in Mathematics **80**, Birkhäuser Verlag, 1984.

[97] J. GLIMM, *Solutions in the large for nonlinear hyperbolic systems of equations*, Comm. Pure Appl. Math. **18** (1965), 697–715.

[98] P. HARTMAN, *Ordinary differential equations*, Classics in Applied Mathematics **38** (2002), Society for Industrial and Applied Mathematics (SIAM), Philadelphia, PA.

[99] M. HAURAY, *On Liouville transport equation with potential in BV_{loc}*, Comm. PDE **29** (2004), 207–217.

[100] M. HAURAY, *On two-dimensional Hamiltonian transport equations with L^p_{loc} coefficients*, Ann. IHP Nonlinear Anal. Non Linéaire **20** (2003), 625–644.

[101] M. HAURAY, C. LE BRIS and P.-L. LIONS, *Deux remarques sur les flots généralisés d'équations différentielles ordinaires*, C. R. Math. Acad. Sci. Paris **344** (2007), no. 12, 759–764.

[102] B. L. KEYFITZ and H. C. KRANZER, *A system of nonstrictly hyperbolic conservation laws arising in elasticity theory*, Arch. Rational Mech. Anal. **72** (1980), 219–241.

[103] S. N. KRUŽKOV, *First order quasilinear equations with several independent variables*, Mat. Sb. (N.S.) **81** (1970), 228–255.

[104] C. LE BRIS and P.-L. LIONS, *Renormalized solutions of some transport equations with partially $W^{1,1}$ velocities and applications*, Annali di Matematica **183** (2003), 97–130.

[105] N. LERNER, *Transport equations with partially BV velocities*, Ann. Sc. Norm. Super. Pisa Cl. Sci. **3** (2004), 681–703.

[106] N. LERNER, *Équations de transport dont les vitesses sont partiellement BV*, Séminaire: Équations aux Dérivées Partielles, Exp. No. X, École Polytech., Palaiseau, 2004.

[107] P.-L. LIONS, *Mathematical topics in fluid mechanics, Vol. I: incompressible models*, Oxford Lecture Series in Mathematics and its applications **3** (1996), Oxford University Press.

[108] P.-L. LIONS, *Mathematical topics in fluid mechanics, Vol. II: compressible models*, Oxford Lecture Series in Mathematics and its applications **10** (1998), Oxford University Press.

[109] P.-L. LIONS, *Sur les équations différentielles ordinaires et les équations de transport*, C. R. Acad. Sci. Paris Sér. I **326** (1998), 833–838.

[110] J. LOTT and C. VILLANI, *Weak curvature conditions and functional inequalities*, J. Funct. Anal. **245** (2007), no. 1, 311–333.

[111] S. MANIGLIA, *Probabilistic representation and uniqueness results for measure-valued solutions of transport equations*, J. Math. Pures Appl. **87** (2007), 601–626.

[112] S. MANIGLIA, *Fine properties and well-posedness of solutions of transport equations and of the 1st order associated ODE*, PhD Thesis, Università di Pisa, 2006.

[113] J. N. MATHER, *Minimal measures*, Comment. Math. Helv. **64** (1989), 375–394.

[114] J. N. MATHER, *Action minimizing invariant measures for positive definite Lagrangian systems*, Math. Z. **207** (1991), 169–207.

[115] E. Y. PANOV, *On strong precompactness of bounded sets of measure-valued solutions of a first order quasilinear equation*, Math. Sb. **186** (1995), 729–740.

[116] G. PETROVA and B. POPOV, *Linear transport equation with discontinuous coefficients*, Comm. PDE **24** (1999), 1849–1873.

[117] F. POUPAUD and M. RASCLE, *Measure solutions to the liner multi-dimensional transport equation with non-smooth coefficients*, Comm. PDE **22** (1997), 337–358.

[118] A. PRATELLI, *Equivalence between some definitions for the optimal transport problem and for the transport density on manifolds*, Ann. Mat. Pura Appl. **184** (2005), 215–238.

[119] S. K. SMIRNOV, *Decomposition of solenoidal vector charges into elementary solenoids and the structure of normal one-dimensional currents*, St. Petersburg Math. J. **5** (1994), 841–867.

[120] E. M. STEIN, *Singular integrals and differentiability properties of functions*, Princeton University Press, 1970.

[121] M. E. TAYLOR, *Partial differential equations. Basic theory.* Texts in Applied Mathematics **23** (1996), Springer-Verlag, New York.

[122] L. TARTAR, *Compensated compactness and applications to partial differential equations*, Research Notes in Mathematics, Nonlinear Analysis and Mechanics, ed. R. J. Knops, vol. **4**, Pitman Press, New York, 1979, 136–211.

[123] R. TEMAM, *Problémes mathématiques en plasticité*, Gauthier-Villars, Paris, 1983.

[124] H. TRIEBEL, *Fractals and Spectra related to Fourier analysis and function spaces*, Monographs in Mathematics **91**, Birkhäuser Verlag, 1997.

[125] H. TRIEBEL and H. WINKELVOSS, *A Fourier analytical characterization of the Hausdorff dimension of a closed set and of related Lebesgue spaces*, Studia Math. **121** (1996), 149–166.

[126] J. I. E. URBAS, *Regularity of generalized solutions of Monge–Ampère equations*, Math. Z. **197** (1988), 365–393.

[127] A. VASSEUR, *Strong traces for solutions of multi-dimensional scalar conservation laws*, Arch. Rational Mech. Anal. **160** (2001), 181–193.

[128] C. VILLANI: *Topics in mass transportation*, Graduate Studies in Mathematics **58** (2004), American Mathematical Society.

[129] C. VILLANI, *Optimal transport: old and new*, Lecture Notes of the 2005 Saint-Flour Summer school.

[130] L. C. YOUNG, *Lectures on the calculus of variations and optimal control theory*, Saunders, 1969.

[131] W. P. ZIEMER, *Weakly differentiable functions. Sobolev spaces and functions of bounded variation*, Graduate Texts in Mathematics **120** (1989), Springer-Verlag.

THESES

This series gathers a selection of outstanding Ph.D. theses defended at the Scuola Normale Superiore since 1992.

Published volumes

1. F. COSTANTINO, *Shadows and Branched Shadows of 3 and 4-Manifolds*, 2005. ISBN 88-7642-154-8

2. S. FRANCAVIGLIA, *Hyperbolicity Equations for Cusped 3-Manifolds and Volume-Rigidity of Representations*, 2005. ISBN 88-7642-167-x

3. E. SINIBALDI, *Implicit Preconditioned Numerical Schemes for the Simulation of Three-Dimensional Barotropic Flows*, 2007.
   ISBN 978-88-7642-310-9

4. F. SANTAMBROGIO, *Variational Problems in Transport Theory with Mass Concentration*, 2007. ISBN 978-88-7642-312-3

5. M. R. BAKHTIARI, *Quantum Gases in Quasi-One-Dimensional Arrays*, 2007. ISBN 978-88-7642-319-2

6. T. SERVI, *On the First-Order Theory of Real Exponentiation*, 2008.
   ISBN 978-88-7642-325-3

7. D. VITTONE, *Submanifolds in Carnot Groups*, 2008.
   ISBN 978-88-7642-327-7

8. A. FIGALLI, *Optimal Transportation and Action-Minimizing Measures*, 2008. ISBN 978-88-7642-330-7

9. A. SARACCO, *Extension Problems in Complex and CR-Geometry*, 2008. ISBN 978-88-7642-338-3

10. L. MANCA, *Kolmogorov Operators in Spaces of Continuous Functions and Equations for Measures*, 2008. ISBN 978-88-7642-336-9

11. M. LELLI, *Solution Structure and Solution Dynamics in Chiral Ytterbium(III) Complexes*, 2009. ISBN 978-88-7642-349-9

12. G. CRIPPA, *The Flow Associated to Weakly Differentiable Vector Fields*, 2009. ISBN 978-88-7642-340-6

Volumes published earlier

H.Y. FUJITA, *Equations de Navier-Stokes stochastiques non homogènes et applications*, 1992.

G. GAMBERINI, *The minimal supersymmetric standard model and its phenomenological implications*, 1993. ISBN 978-88-7642-274-4

C. DE FABRITIIS, *Actions of Holomorphic Maps on Spaces of Holomorphic Functions*, 1994. ISBN 978-88-7642-275-1

C. PETRONIO, *Standard Spines and 3-Manifolds*, 1995. ISBN 978-88-7642-256-0

I. DAMIANI, *Untwisted Affine Quantum Algebras: the Highest Coefficient of* det H_η *and the Center at Odd Roots of 1*, 1996. ISBN 978-88-7642-285-0

M. MANETTI, *Degenerations of Algebraic Surfaces and Applications to Moduli Problems*, 1996. ISBN 978-88-7642-277-5

F. CEI, *Search for Neutrinos from Stellar Gravitational Collapse with the MACRO Experiment at Gran Sasso*, 1996. ISBN 978-88-7642-284-3

A. SHLAPUNOV, *Green's Integrals and Their Applications to Elliptic Systems*, 1996. ISBN 978-88-7642-270-6

R. TAURASO, *Periodic Points for Expanding Maps and for Their Extensions*, 1996. ISBN 978-88-7642-271-3

Y. BOZZI, *A study on the activity-dependent expression of neurotrophic factors in the rat visual system*, 1997. ISBN 978-88-7642-272-0

M.L. CHIOFALO, *Screening effects in bipolaron theory and high-temperature superconductivity*, 1997. ISBN 978-88-7642-279-9

D.M. CARLUCCI, *On Spin Glass Theory Beyond Mean Field*, 1998. ISBN 978-88-7642-276-8

G. LENZI, *The MU-calculus and the Hierarchy Problem*, 1998. ISBN 978-88-7642-283-6

R. SCOGNAMILLO, *Principal G-bundles and abelian varieties: the Hitchin system*, 1998. ISBN 978-88-7642-281-2

G. ASCOLI, *Biochemical and spectroscopic characterization of CP20, a protein involved in synaptic plasticity mechanism*, 1998. ISBN 978-88-7642-273-7

F. PISTOLESI, *Evolution from BCS Superconductivity to Bose-Einstein Condensation and Infrared Behavior of the Bosonic Limit*, 1998. ISBN 978-88-7642-282-9

L. PILO, *Chern-Simons Field Theory and Invariants of 3-Manifolds*, 1999. ISBN 978-88-7642-278-2

P. ASCHIERI, *On the Geometry of Inhomogeneous Quantum Groups*, 1999. ISBN 978-88-7642-261-4

S. CONTI, *Ground state properties and excitation spectrum of correlated electron systems*, 1999. ISBN 978-88-7642-269-0

G. GAIFFI, *De Concini-Procesi models of arrangements and symmetric group actions*, 1999. ISBN 978-88-7642-289-8

N. DONATO, *Search for neutrino oscillations in a long baseline experiment at the Chooz nuclear reactors*, 1999. ISBN 978-88-7642-288-1

R. CHIRIVÌ, *LS algebras and Schubert varieties*, 2003.
ISBN 978-88-7642-287-4

V. MAGNANI, *Elements of Geometric Measure Theory on Sub-Riemannian Groups*, 2003. ISBN 88-7642-152-1

F.M. ROSSI, *A Study on Nerve Growth Factor (NGF) Receptor Expression in the Rat Visual Cortex: Possible Sites and Mechanisms of NGF Action in Cortical Plasticity*, 2004. ISBN 978-88-7642-280-5

G. PINTACUDA, *NMR and NIR-CD of Lanthanide Complexes*, 2004.
ISBN 88-7642-143-2

Fotocomposizione "CompoMat" Loc. Braccone, 02040 Configni (RI) Italy
Finito di stampare nel mese di febbraio 2009
dalla CSR srl, Via di Pietralata, 157, 00158 Roma